M.Eng. **René Schwarz**

DEVELOPMENT OF AN ILLUMINATION SIMULATION SOFTWARE FOR THE MOON'S SURFACE

An approach to illumination direction estimation on pictures of solid planetary surfaces with a significant number of craters.

BASED UPON THE ORIGINAL EXAMINATION VERSION

Cover Images
- Moon photograph: *S103-E-5037 (December 21, 1999)* — *Astronauts aboard the Space Shuttle Discovery recorded this rarely seen phenomenon of the full Moon partially obscured by the atmosphere of Earth. The image was recorded with an electronic still camera at 15:15:15 GMT, Dec. 21, 1999.* © NASA, available at http://spaceflight.nasa.gov/gallery/images/shuttle/sts-103/html/s103e5037.html.
- Moon surface tile: *Simulated map of safe landing areas around a large lunar crater.* © NASA/Goddard Scientific Visualization Studio, available at http://www.nasa.gov/images/content/271355main_safeonly_print_jpg.jpg.

Meinen Großeltern,

Rosa Maria und Konrad Siermann,
Gerda Schwarz

meinen Eltern,

Angela und Sven Schwarz,

die mich mein Leben lang nach Kräften in meinen Bestrebungen unterstützt haben.

DEVELOPMENT OF AN ILLUMINATION SIMULATION SOFTWARE FOR THE MOON'S SURFACE

An approach to illumination direction estimation on pictures of solid planetary surfaces with a significant number of craters.

— Master's Thesis —

BY

Mr. B.Eng. **René Schwarz** (Matriculation Number 17288)
né Siermann, born on April 15th, 1987 in Merseburg, Germany

SUBMITTED TO THE DEPARTMENT OF COMPUTER SCIENCE AND COMMUNICATION SYSTEMS AND THE INSTITUTE OF SPACE SYSTEMS (GERMAN AEROSPACE CENTER, DLR) IN PARTIAL FULFILLMENT OF THE REQUIREMENTS FOR THE DEGREE OF

MASTER OF ENGINEERING (M.ENG.)
IN COMPUTER SCIENCE/ARTIFICIAL INTELLIGENCE

AT THE
MERSEBURG UNIVERSITY OF APPLIED SCIENCES

ON THIS 9TH DAY OF APRIL, 2012.

SUPERVISORS

GERMAN AEROSPACE CENTER (DLR)	MERSEBURG UNIVERSITY OF APPLIED SCIENCES
Institute of Space Systems	Department of Computer Science and Communication Systems
Dipl. Math.-Techn. **Bolko Maass**	Prof. Dr. rer. nat. **Hartmut Kröner**

ENTWICKLUNG EINER SOFTWARE ZUR SIMULATION DER OBERFLÄCHENBELEUCHTUNG DES MONDES

Ein Zugangsweg zur Schätzung der Beleuchtungsrichtung auf Bildern planetarer Oberflächen mit einer signifikanten Anzahl von Kratern.

— MASTERTHESIS —

VON

Herrn B.Eng. **René Schwarz** (Matr.-Nr. 17288)
geb. Siermann, geboren am 15. April 1987 in Merseburg

VORGELEGT DEM FACHBEREICH INFORMATIK UND KOMMUNIKATIONSSYSTEME UND DEM INSTITUT FÜR RAUMFAHRTSYSTEME (DEUTSCHES ZENTRUM FÜR LUFT- UND RAUMFAHRT, DLR) ZUR ERLANGUNG DES AKADEMISCHEN GRADES EINES

**MASTER OF ENGINEERING (M.ENG.)
IN INFORMATIK/KÜNSTLICHE INTELLIGENZ**

AN DER
HOCHSCHULE MERSEBURG

AM 09. APRIL 2012.

MENTOREN

DEUTSCHES ZENTRUM FÜR LUFT- UND RAUMFAHRT (DLR) Institut für Raumfahrtsysteme	HOCHSCHULE MERSEBURG Fachbereich Informatik und Kommunikationssysteme
Dipl. Math.-Techn. **Bolko Maass**	Prof. Dr. rer. nat. **Hartmut Kröner**

M.Eng. René Schwarz
Merseburg University of Applied Sciences
German Aerospace Center (DLR), Institute of Space Systems, Bremen

e-mail: mail@rene-schwarz.com · web: http://www.rene-schwarz.com

Citation Proposal

Schwarz, René: *Development of an illumination simulation software for the Moon's surface: An approach to illumination direction estimation on pictures of solid planetary surfaces with a significant number of craters*. Master's Thesis, Merseburg University of Applied Sciences, German Aerospace Center (DLR). Books on Demand, Norderstedt, Germany, 2012. ISBN 978-3-8482-1628-4.

BibTeX entry:
```
@book{schwarz-rene-2012-illumination,
    author = {Schwarz, Ren\'e},
    title = {Development of an illumination simulation software for the Moon's surface:
             An approach to illumination direction estimation on pictures of solid planetary
             surfaces with a significant number of craters},
    publisher = {Books on Demand, Norderstedt, Germany},
    note = {Master's Thesis, Merseburg University of Applied Sciences,
            German Aerospace Center (DLR)},
    year = {2012},
    isbn = {978-3-8482-1628-4},
    url = {http://go.rene-schwarz.com/masters-thesis}
}
```

Bibliografische Information der Deutschen Nationalbibliothek

Die Deutsche Nationalbibliothek verzeichnet diese Publikation in der Deutschen Nationalbibliografie; detaillierte bibliografische Daten sind im Internet über http://dnb.d-nb.de abrufbar.

Copyright ©2012 M.Eng. René Schwarz (rene-schwarz.com), unless otherwise stated.

This book is subject to the conditions of the Creative Commons Attribution-NonCommercial 3.0 Unported License (CC BY-NC 3.0); it can be obtained in a digital version (free of charge) as well as in form of a printed paperback copy. Please attribute/cite this work in the way specified above. If you want to use this book or parts of it for commercial purposes, please contact the author via e-mail.

Any trademarks, service marks, product names or named features are assumed to be the property of their respective owners; they are used throughout this book in an editorial fashion only. There is no implied endorsement/relationship.

Although the author has used his best efforts in preparing this book, he assumes no responsibility for errors or omissions, especially the information is represented here without any warranty; without even the implied warranty of merchantability or fitness for a particular purpose. Other licenses can apply for some contents; especially source code written by the author is released subject to the conditions of the GNU General Public License (GPL), version 2 or later, unless otherwise stated.

Document Preparation and Layout:	M.Eng. René Schwarz
Illustrations and Artwork:	Matthias Kopsch (unless otherwise stated)
Printing and Publishing:	BoD – Books on Demand GmbH, Norderstedt.

Printed in Germany. Typeset using X₃LATEX as part of the MiKTEX TEX distribution.

Digital version (PDF): http://go.rene-schwarz.com/masters-thesis
Paperback copy ISBN: 978-3-8482-1628-4

Abstract

The exploration of the solar system over the last decades has broadened our knowledge and understanding of the universe and our place in it. Great scientific and technological achievements have been made, allowing us to study faraway places in the solar system. The world's space agencies are now facing a new era of continuing space exploration in the 21st century, expanding permanent human presence beyond low Earth orbit for the first time. To pursue this goal, the development of advanced technologies is more urgent than ever before.

One key technology for future human and robotic missions to places distant from Earth will be a system for autonomous navigation and landing of spacecraft, since nowadays navigation systems rely on Earth-based navigation techniques (tracking, trajectory modeling, commanding). A promising approach involves optical navigation technologies, which can operate completely independently of Earth-based support, allowing a surface-relative navigation and landing on celestial bodies without human intervention.

The German Aerospace Center (DLR) is developing a new, holistic optical navigation system for all stages of an approach and landing procedure within the ATON project (Autonomous Terrain-based Optical Navigation). The central feature of this new navigation system is its landmark-based navigation. Commonly, craters are used as landmarks, as they exhibit very characteristic shapes and they are stable over the long term with respect to shape, structure and positioning. However, the flawless perception of these surface features by computers is a non-trivial task.

A new edge-free, scale-, pose- and illumination-invariant crater detection algorithm is developed for ATON, which will do away with the limitations of current algorithms. To promote further development, the possibility of generating realistic surface images of celestial bodies with a significant number of craters and with well-known local illumination conditions is essential, as well as a technique for estimating the local illumination direction on these images. To date, no software exists to generate artificial renderings of realistically illuminated planetary surfaces while determining the local solar illumination direction.

Having said this, the objective of this thesis is the development of a surface illumination simulation software for solid planetary surfaces with a significant number of craters, whereas all work has been done in the context of the Moon. The thesis work has led to the development of the *Moon Surface Illumination Simulation Framework* (MSISF), which is the first software known to produce realistic renderings of the entire Moon's surface from virtually every viewpoint, while simultaneously generating machine-readable information regarding the exactly known parameters for the environmental conditions, such as the local solar illumination angle for every pixel of a rendering showing a point on the Moon's surface.

To produce its renderings, the MSISF maintains a global digital elevation model (DEM) of the Moon, using the latest data sets from the ongoing NASA *Lunar Reconnaissance Orbiter* (LRO) mission. The MSISF has also demonstrated its ability to not only produce single renderings, but also whole series of renderings corresponding to a virtual flight trajectory or landing on the Moon. This thesis shows how these renderings will be produced and how they will be suitable for the development and testing of new optical navigation algorithms. The MSISF can also be modified for the rendering of other celestial bodies.

With the MSISF, a basis has been established for the further development of the new DLR crater detection algorithm as well as for the illuminance flow estimation on pictures of solid planetary surfaces.

Keywords: Moon, 3D Model, 3D Rendering, Illumination Simulation, Illumination Direction Estimation, Illuminance Flow Estimation, Local Solar Illumination Angle, Terrain-Relative Navigation, Optical Navigation, Digital Elevation Model, Digital Terrain Model, Digital Surface Model, DELAUNAY Triangulation, Crater Detection, Topography, MySQL Spatial Database, Surface Pattern, Ray Tracing, Dynamical Surface Pattern Selection Algorithm, Moon Surface Illumination Simulation Framework, MSISF, C#, PHP, XML, SPICE, NASA, DLR

Abstract (Deutsch)

Die Erkundung unseres Sonnensystems in den vergangenen Jahrzehnten erweiterte unser Wissen und Verständnis des Universums und von unserem Platz darin; großartige wissenschaftliche und technische Errungenschaften gestatten uns das Studium weit entfernter Orte im Sonnensystem. Weltweit sehen Raumfahrtagenturen nun einer neuen Ära der Erforschung des Sonnensystems im 21. Jahrhundert entgegen, in der die menschliche Präzens im Weltall erstmals dauerhaft außerhalb eines niedrigen Erdorbits ausgeweitet werden soll. Um dieses Ziel erfolgreich verfolgen zu können, drängt die Entwicklung neuer Technologien mehr als je zuvor.

Eine Schlüsseltechnologie für künftige bemannte und unbemannte Missionen zu weit von der Erde entfernten Zielen wird die Fähigkeit zur autonomen, sicheren Navigation und Landung von Raumfahrzeugen sein, da momentane Navigationssysteme auf erdgebundenen Techniken (Bahnverfolgung und -modellierung, Steuerung) basieren. Ein vielversprechender Ansatz dazu sind optische Navigationstechnologien, die komplett unabhängig von der Erde arbeiten und damit eine Oberflächen-relative Navigation und Landung ohne menschliches Eingreifen erlauben.

Das Deutsche Zentrum für Luft- und Raumfahrt (DLR) entwickelt derzeit ein neuartiges, ganzheitliches, optisches Navigationssystem für alle Phasen der Annäherung an einen Himmelskörper sowie der anschließenden Landeprozedur innerhalb des ATON-Projekts (Autonomous Terrain-based Optical Navigation). Den Kern dieses neuen Navigationssystems bildet die Navigation anhand von Orientierungspunkten auf Himmelskörpern. Üblicherweise werden Krater als solche Orientierungspunkte verwendet, da sie charakteristische Merkmale aufweisen, die langzeitstabil in Bezug auf Form, Struktur und Anordnung sind. Nichtsdestotrotz bleibt die reibungslose Kratererkennung durch Computer nach wie vor eine nicht-triviale Aufgabe.

Für ATON wird zudem momentan ein Algorithmus zur Kratererkennung entwickelt, der ohne Kantenerkennung auskommt sowie unabhängig von Skalierung, Blick- und Beleuchtungswinkel ist und somit die Beschränkungen der derzeitigen Verfahren durchbricht. Für die weitere Entwicklung ist eine Möglichkeit zur Erzeugung von realistischen Bildern (sogenannte Renderings) von Himmelskörpern mit einer signifikanten Anzahl von Kratern, zu denen die lokale Beleuchtungsrichtung bekannt ist, maßgeblich, ebenso wie die Entwicklung von geeigneten Verfahren, mit denen eine Schätzung der lokalen Beleuchtungsrichtung auf solchen Bildern ermöglicht wird. Bislang ist jedoch keine Software bekannt, die in der Lage ist, derartige Bilder unter direkter Angabe der lokalen Beleuchtungsrichtung zu erzeugen.

Daher ist die Hauptaufgabe dieser Thesis die Entwicklung einer Software zur Simulation der Beleuchtung fester, planetarer Oberflächen mit einer signifikanten Anzahl von Kratern, wobei die gesamte Arbeit im Kontext des Erdmondes als Vertreter dieser Klasse von Himmelskörpern ausgeführt worden ist. Die praktischen Arbeiten im Rahmen der Thesis führten zur Entwicklung des *Moon Surface Illumination Simulation Framework* (MSISF), welches die derzeit erste Software darstellt, die realistische Renderings der gesamten Mondoberfläche von praktisch jedem Beobachtungsort im Orbit des Mondes erzeugen kann und dabei gleichzeitig computerlesbare Metadaten zu den exakten Umgebungsbedingungen erzeugt, wie zum Beispiel den lokalen Beleuchtungswinkel der Sonne für jeden Pixel eines erzeugten Bildes, der einen Punkt der Mondoberfläche darstellt.

Um realistische Renderings erzeugen zu können, bedient sich das MSISF an einem globalen, digitalen Oberflächenmodell des Mondes, das auf den neuesten Datensätzen der laufenden *Lunar Reconnaissance Orbiter* (LRO) Mission der NASA basiert. Im Laufe der Arbeit zeigte sich zudem, dass das MSISF nicht nur einzelne Bilder, sondern ganze Bildfolgen einer kompletten virtuellen Flugbahn oder Landung auf dem Mond effizient erzeugen kann.

Mit dem MSISF wurde sowohl ein Grundstein für die weitere Entwicklung des neuen Kratererkennungsalgorithmus des DLR gelegt, als auch für weitere Arbeiten auf dem Gebiet der Schätzung der lokalen Beleuchtungsrichtung auf Bildern fester, planetarer Oberflächen.

Keywords: Mond, 3D-Modell, 3D-Rendering, Beleuchtungssimulation, Schätzung der Beleuchtungsrichtung, Oberflächen-relative Navigation, Optische Navigation, Digitales Geländemodell, Digitales Oberflächenmodel, DELAUNAY-Triangulation, Kratererkennung, Topografie, MySQL Spatial Extensions, Surface Pattern, Raytracing, dynamische Patternauswahl, lokaler Beleuchtungswinkel, lokale Beleuchtungsrichtung, Moon Surface Illumination Simulation Framework, MSISF, C#, PHP, XML, SPICE, NASA, DLR

Table of Contents

Preface 1

List of Abbreviations 3

Notation Overview 7

1 Thesis Background and Scope 13
    1.1 Future Challenges of Space Exploration Missions 13
    1.2 Necessity of New Navigation Technologies 16
    1.3 Ambitions of the German Aerospace Center (DLR) 17
    1.4 Thesis' Contribution to Novel Navigation Systems 24

2 Introducing the Moon Surface Illumination Simulation Framework (MSISF) .. 29
    2.1 General Concept 29
    2.2 Development Milestones 30
        2.2.1 Reference Frame and Selenographic Coordinates 30
        2.2.2 Lunar Topography Data Acquisition 32
        2.2.3 Database Creation, Data Import and Conditioning 32
        2.2.4 Surface Pattern Generation 32
        2.2.5 Surface Pattern Selection, Assembling and Rendering 33
        2.2.6 Output of the Local Solar Illumination Angle 33
        2.2.7 Result Discussion 33
    2.3 Preparatory Remarks 34
        2.3.1 Software and Programming Languages Used, Development Environment . 34
        2.3.2 MSIS User Interface, Definition of Inputs and Outputs 36
        2.3.3 MSISF File System Layout 37
        2.3.4 MSISF Deployment and Installation 38

3 Theoretical Foundations 39
    3.1 The Mean Earth/Polar Axis Reference Frame 39
    3.2 Derivation of a Spherical Coordinate System for the Moon 40

4 Creating a Global Lunar Topographic Database — 45
- 4.1 Overview of Available Lunar Topographic Data 45
- 4.2 Data from NASA's Lunar Orbiter Laser Altimeter (LOLA) 47
- 4.3 LOLA Data Import and Conditioning 51
  - 4.3.1 Overall Concept 51
  - 4.3.2 MySQL Database Design, Query Optimization and Commitment of the MySQL Server Configuration to Database Performance 52
  - 4.3.3 Importing the LOLA Data into the MySQL Database 59

5 Surface Pattern Generation Process — 67
- 5.1 Anatomy of a Surface Pattern 67
- 5.2 Surface Pattern Generation Process 72
- 5.3 Storage of the Surface Patterns and POV-Ray Mesh Compilation 74

6 Moon Surface Illumination Simulator (MSIS) — 79
- 6.1 Software Architecture 79
- 6.2 Selected Components of the MSIS 81
  - 6.2.1 Determination of the Sun's Position Using NASA NAIF SPICE 81
  - 6.2.2 Position Calculation Using a Set of Keplerian Orbit Elements 84
  - 6.2.3 State Vector Conversion to Keplerian Elements 87
  - 6.2.4 Time Calculations 90
- 6.3 User Interface and MSIS Invocation 92
  - 6.3.1 General Information 92
  - 6.3.2 Batch File Operation Mode 92
  - 6.3.3 Fixed State Operation Mode 94
  - 6.3.4 Keplerian Elements Operation Mode 94
  - 6.3.5 State Vectors Operation Mode 94
- 6.4 Example Usage 94

7 Spacecraft Orientation and Rotation Model Using Quaternions — 97
- 7.1 Introduction to Quaternions and Spatial Rotation 97
- 7.2 Spacecraft Orientation and Rotation Model 100

8 Dynamical Surface Pattern Selection — 103
- 8.1 Ray Tracing with a Sphere 104
- 8.2 Camera Geometry 105
- 8.3 MSIS Implementation 109
- 8.4 Drawbacks of this Method 112

9 XML Meta Rendering Information and Rendering Annotations — 115
- 9.1 Definition of the MSIS Output ... 115
- 9.2 Structure of the XML Meta Information File ... 116
- 9.3 Determination of the Local Solar Illumination Angle ... 122

10 Results, Discussion and Conclusion — 135
- 10.1 Synopsis of the Work Results ... 135
- 10.2 Suggestions for Improvements and Future Work ... 137
  - 10.2.1 Performance Optimization ... 137
  - 10.2.2 Improvement of the Topography Database ... 138
  - 10.2.3 Graphical User Interface ... 138
  - 10.2.4 Rendering Parallelization/Distributed Rendering ... 138
  - 10.2.5 Utilization of the MSISF for Other Celestial Bodies ... 138
  - 10.2.6 Real-Time Video Preparation ... 139
  - 10.2.7 Compile POV-Ray as Windows Command-Line Tool ... 139
  - 10.2.8 INI Settings ... 139
  - 10.2.9 Original Implementation of the 2D DELAUNAY Triangulation ... 139
  - 10.2.10 Compensate the Drawbacks of the DSPSA ... 140
- 10.3 Construction Progress of TRON ... 140
- 10.4 MSISF Application at DLR ... 142

Appendices

A MSIS User Interface Specification — 145

B Code Listings — 151
- B.1 MySQL Server Instance Configuration ... 151
- B.2 MSISRendering XML Document Type Definition (DTD) ... 156
- B.3 LDEM Import Script ... 158
- B.4 Pattern Generation Script ... 165

Bibliography — 173
- Astrodynamics/Celestial Mechanics ... 173
- Computer Vision (General) ... 175
- Illuminance Flow Estimation ... 175
- Mathematics/Physics in General, Numerical Analysis and Computational Science ... 179

Scientific Tables/Conventions, Works of Reference, Algorithms 180
Spacecraft Engineering 181
Other Topics 181

Listings 187
List of Figures 187
List of Tables 190
List of Code Listings 191

Alphabetical Index 193

Preface

Finally, after a year of productive and challenging labor, I am grateful for the chance to publish the results of this fascinating research I have been permitted to work on with the joint support of the Institute of Space Systems of the German Aerospace Center (DLR) in Bremen and the Merseburg University of Applied Sciences. During this time, I have gained a deeper insight into the fields of mathematics, computer graphics and astrodynamics. I have endeavored to create a thesis that will prove its worth by being of benefit to future research.

This thesis would not have been possible without the outstanding cooperative support of colleagues, followers, mentors and friends. I would like to give my sincere thanks to my supervisors, Bolko Maass from the Institute of Space Systems of the German Aerospace Center (DLR) and Professor Hartmut Kröner from the Merseburg University of Applied Sciences, who gave me the opportunity to work on such an attractive subject and enabled the essential conditions for the creative mental work on it — and, above all things, for their unprecedented support and their personal and professional counseling throughout the preparation of this thesis.

I would further like to thank my grandparents, parents and especially Matthias Kopsch for the unconditional succor in areas where only they could be on hand with help and advice. A special thank you is addressed to Yolande McLean, Canada, for her support in patiently proofreading all the drafts of this thesis.

Additionally, this thesis would not have been possible in the first place without the thousands of people creating, supporting and enhancing the myriad of free and/or open source software, namely the Apache HTTPD Webserver, Celestia, doxygen, GIMP, Inkscape, IrfanView, FileZilla, Firefox, MediaWiki, MeshLab, MiKTeX (X$_\exists$L^{A}T$_E$X), MySQL, Notepad++, PHP, phpMyAdmin, POV-Ray, SPICE, SVN, TeXnicCenter, Thunderbird, TortoiseSVN, VLC, WinMerge, yEd and all the ones I have forgotten.

Finally, I would like to thank the National Aeronautics and Space Administration (NASA) for producing and providing high quality data and services, especially — but not limited to — regarding the Lunar Reconnaissance Orbiter (LRO) mission.

A Note on the License

I understand knowledge as one essence of human life, and therefore I hold the opinion that all scientific achievements should be free and accessible for the scientific community and all interested people, ensuring the development of human knowledge regardless of financial or social circumstances, ethnicity or technological capabilities.

Thus, permission is hereby granted, free of charge, to use, copy, modify, merge or share this book subject to the conditions of the Creative Commons Attribution-NonCommercial 3.0 Unported License (CC BY-NC 3.0)[1]. This book can be obtained in a digital version (PDF) from http://go.rene-schwarz.com/masters-thesis as well as in the form of a printed paperback copy (see imprint for ISBN/publisher).

Please note that other licenses may apply for some contents of this book; in particular, all source code written by the author is subject to the conditions of the GNU General Public License (GPL), version 2 or later, unless otherwise stated. Material with an external copyright notice is reproduced here with friendly permission; however, please note that one must obtain a new license for any work derived from this book containing those elements from the respective copyright owners.

Online Supplemental Material

Additional, supplemental material for this thesis can be found online at http://go.rene-schwarz.com/masters-thesis, including videos of rendering sequences prepared during the thesis work. Furthermore, the MSISF (the software developed during this thesis) will be available for download at this place, along with errata or additional remarks, if applicable. This website will be extended and revised from time to time.

---

[1] License information: http://creativecommons.org/licenses/by-nc/3.0/; please attribute/cite this work in the way specified in the imprint, section "citation proposal". If you want to use this book or parts of it for commercial purposes, please contact me via e-mail at mail@rene-schwarz.com.

List of Abbreviations

ACID	*Atomicity, Consistency, Isolation and Durability*, a concept for the reliable processing of database transactions
ATON	*Autonomous Terrain-based Optical Navigation*, a project by the GNC devision of the DLR/IRS for the development of a novel optical navigation system
AU	*Astronomical Unit*, a length unit based on the mean distance between Sun and Earth
Caltech	*California Institute of Technology*, a well-known U.S. university located in Pasadena, California
COM	*Center of Mass*, a point of a body in space, at which the entire body's mass maybe assumed to be concentrated
CPU	*Central Processing Unit*, the main processor of a computer, which is able to execute programs
CSV	*Comma-Separated Values*, a file format for storing tabular data
CxP	*Constellation Program*, a former NASA program for the human exploration of Mars, canceled in 2010
DEM	*Digital Elevation Model*, a digital 3D surface model of a celestial body
DLL	*Dynamic-Link Library*, a compiled, shared library to be used in other applications running on Microsoft Windows
DLR	*Deutsches Zentrum für Luft- und Raumfahrt e. V.* (English: German Aerospace Center), Germany's space agency
DLR/IRS	*Institut für Raumfahrtsysteme* (English: Institute of Space Systems), an institute of the DLR in Bremen, Germany
DOF	*Degrees of Freedom*, the number of independent translations/rotations a mechanical system can perform
DOI	*Descent Orbit Injection*, an orbital maneuver to bring a spacecraft out of a stable parking orbit in order to achieve an unpowered descent to the surface of the orbited body
DTD	*Document Type Definition*, a file describing the syntax and semantics of a XML file

DSPSA	*Dynamical Surface Pattern Selection Algorithm*, an algorithm developed in this thesis to dynamically select needed surface patterns for renderings (see chapter 8)
DSPSE	*Deep Space Program Science Experiment*, the official name of NASA's Clementine Mission
ESA	*European Space Agency*, the European intergovernmental organization for space exploration
ET	*Ephemeris Time*, a system for time measurement often used in conjunction with ephemeris
FOV	*Field of View*, an angular measure for the visible area seen from an optical instrument
FPGA	*Field-Programmable Gate Array*, a re-configurable integrated circuit
GNC	*Guidance, Navigation and Control*, a field of engineering for the development of systems controlling the movement of spacecraft
GNU	*GNU's not UNIX*, a recursive acronym for a computer operating system
GPL	*GNU General Public License*, a free software license
GPU	*Graphics Processing Unit*, the processor of a graphics adapter
GUI	*Graphical User Interface*, a software user interface using graphical elements instead of text commands
HA	*Hazard Avoidance*, a general term for all methods and techniques to protect a spacecraft from hazards during landing operations
HDD	*Hard Disk Drive*, a storage device in computers
HiL	*Hardware-in-the-Loop*, a real-time simulation technique including physical sensors or actuators
IAG	*International Association of Geodesy*, an association of the IUGG
IAU	*International Astronomical Union*, an internationally recognized association of professional astronomers and researchers; authority for, among others, astronomical naming conventions and planetary nomenclature
IDE	*Integrated Development Environment*, a software development application suite
INI	*Initialization File*, a file storing application settings
IUGG	*International Union of Geodesy and Geophysics*, an international organization for Earth sciences and studies
JAXA	*Japan Aerospace Exploration Agency* (Japanese: 独立行政法人宇宙航空研究開発機構), the Japanese space agency
JD	*Julian Date*, a system for measuring time used in astronomy

JPL	*Jet Propulsion Laboratory*, a well-known NASA facility managed by the California Institute of Technology, Pasadena, California
LALT	*Laser Altimeter*, an instrument aboard the JAXA SELENE mission
LAN	*Longitude of the Ascending Node*, an angle used for specifying the orbit of an object in space with relation to a reference frame
LDEM	*Lunar Digital Elevation Model*, a DEM of the Moon (see definition of DEM before)
LEO	*low Earth orbit*, a geocentric orbit with apogee and perigee below 2 000 km altitude above mean sea level
LIDAR	*Light Detection and Ranging*, a technique for measuring distance to a target
LOLA	*Lunar Orbiter Laser Altimeter*, an instrument of the Lunar Reconnaissance Orbiter
LRO	*Lunar Reconnaissance Orbiter*, a NASA mission to the Moon launched in 2009
MBR	*Minimum Bounding Rectangle*, the smallest possible axes-parallel rectangle enclosing a defined set of points or objects in a two-dimensional space
MCR	*MATLAB Compiler Runtime*, a runtime enviroment for compiled MATLAB scripts and applications
ME/PA	*Mean Earth/Polar Axis*, a reference frame for the Moon, recommended by the IAU
MJD	*Modified Julian Date*, the result of the subtraction of a Julian date and the number 2 400 000.5
MMR	*Mean Moon Radius*, the volumetric mean radius (as defined by the IUGG) of the Moon
MSIS	*Moon Surface Illumination Simulator*, the nucleus of the MSISF producing renderings of the Moon's surface with variable illumination and observation conditions
MSISF	*Moon Surface Illumination Simulation Framework*, the software developed during this thesis for the illumination simulation of the Moon's surface
NAIF	*Navigation and Ancillary Information Facility*, a small group of NASA people responsible for the development of SPICE located at the JPL
NASA	*National Aeronautics and Space Administration*, the U.S. space agency
PDI	*Powered Descent Initiate*, the initiation of a braking action subsequent to a DOI, reducing the relative speed of a spacecraft to the orbited body's surface
PDS	*Planetary Data System*, a data distribution system for NASA planetary missions
PHP	originally *Personal Home Page*, now *Hypertext Preprocessor*, a software for generating dynamical websites on webservers

PNG	*Portable Network Graphics*, a file format for storing raster graphics
RAM	*Random-Access Memory*, a high-speed computer memory for storing data of program executions
RGB	*Red Green Blue*, a color system
SELENE	*Selenological and Engineering Explorer*, a lunar orbiter by JAXA, which is also known as Kaguya
SDL	*Scene Description Language*, a programming language used for the description of rendering scenes in POV-Ray
SPICE	*Spacecraft Planet Instrument C-matrix Events*, a NASA NAIF toolkit for the computation of geometric and astrodynamical information of objects in space; available for C, Fortran, IDL and MATLAB
SQL	*Structured Query Language*, a programming language for querying/commanding a relational database system
SSD	*Solid-State Disk*, a data storage device consisting of flash memory intended as successor of conventional hard disk drives
TC	*Terrain Camera*, an optical instrument aboard the JAXA SELENE mission
TIN	*Triangulated Irregular Network*, a digital surface representation using faces of interconnected triangles, which have been produced by triangulation of a point cloud (f.e. with a DELAUNAY Triangulation)
TRN	*Terrain-Relative Navigation*, a navigation technique that uses the comparision of an a-priori reference map and in-situ measurements to gain position and orientation information about spacecraft during approach to and descent on celestial bodies
TRON	*Testbed for Robotic Optical Navigation*, a laboratory facility for tests and evaluations within the DLR ATON project
UT1	*Universal Time 1*, a system for time measurement based on Earth's rotation
UTC	*Coordinated Universal Time*, a system for measuring time, which is the world's primary time standard
WPF	*Windows Presentation Foundation*, a software for generating GUIs in Windows
XML	*Extensible Markup Language*, a file format for storing machine-readable, structured data

Notation Overview

General Notation Style

- All **italic Latin and Greek symbols**, both minuscules and majuscules, represent **scalar quantities**, which can be variables, physical symbols or constants.
  Examples: $a, b, c, A, B, C, \alpha, \beta, \Xi, \Psi, \ldots$

- **Bold Latin minuscules** represent **vector quantities**. The components of a vector $\mathbf{x}$ are written as the italic symbol of $\mathbf{x}$ (since they are scalars) with running indices. So the first component of vector $\mathbf{x}$ is x_1, the second x_2 and so on until the last component x_n. Vectors are always column vectors (row vectors are written as the transpose of column vectors).
  Examples: $\mathbf{a}, \mathbf{b}, \ldots$

- **Bold Latin majuscules** represent $m \times n$-**matrices**, where m is the number of rows and n the number of columns. One matrix component is written as the italic Latin minuscule of the bold matrix symbol, indexed by two numbers i and j, where the row of the component is given by i and the column by j.
  Examples: $\mathbf{A}, \mathbf{B}, \ldots$
  Matrix components:
  $$\mathbf{X} = \begin{pmatrix} x_{1,1} & x_{1,2} & \cdots & x_{1,n} \\ x_{2,1} & x_{2,2} & \cdots & x_{2,n} \\ \vdots & \vdots & \ddots & \vdots \\ x_{m,1} & x_{m,2} & \cdots & x_{m,n} \end{pmatrix}$$

- **Latin blackboard majuscules** denote **sets of numbers**.
  Examples: $\mathbb{R}$ for the set of real numbers, $\mathbb{H}$ for the set of quaternions

- **Upright Latin symbols and strings**
  - imply **functions, operators or constants** with a traditional meaning
    Examples: differtial operator d, EULER's number e, imaginary units i, j, k, sine function sin

- define **new operators**

  Example: function $\text{PixelPos}(x, y)$ to get the spatial position of one pixel

- are **part of indices as abbreviations or complete words**

  Example: camera position $\mathbf{c}_{\text{pos}}$

- are **variables in a textual equation**

  Example: direction $= \mathbf{c}_{\text{pointing}}$

■ All **other symbols** have **special meanings** (see tables following) and are written in a non-standard way because of tradition, practice or importance. These symbols are introduced at their time of appearance.

■ **Numbers with digits in parentheses**, e.g. in $6.67428(67) \cdot 10^{-11}$ (the value of the NEWTONian constant of gravitation), are a common way to state the **uncertainty**; this is a short notation for $(6.67428 \pm 0.0000067) \cdot 10^{-11}$.

Notation	Meaning	Note
$\stackrel{!}{=}, \stackrel{\text{def.}}{=\!=}$	definition symbols	
$\langle \mathbf{x}, \mathbf{y} \rangle$	dot product of vector $\mathbf{x}$ and $\mathbf{y}$	$\langle \mathbf{x}, \mathbf{y} \rangle \stackrel{!}{=} \sum_{i=1}^{n} x_i y_i$
$\|\mathbf{x}\|$	EUCLIDean norm (2-Norm) of vector $\mathbf{x}$	$\|\mathbf{x}\| \stackrel{!}{=} \sqrt{x_1^2 + x_2^2 + \ldots x_n^2}$ $\|\mathbf{x}\| \stackrel{!}{=} \langle \mathbf{x}, \mathbf{x} \rangle$
$\mathbf{a} \times \mathbf{b}$	vector cross product of vectors $\mathbf{a}$ and $\mathbf{b}$	
$\mathbf{M} \cdot \mathbf{N}$ or $\mathbf{MN}$	matrix multiplication of matrices $\mathbf{M}$ and $\mathbf{N}$	
$q = [q_0, \mathbf{q}]$	quaternion (in general) with real part q_0 and imaginary vector part $\mathbf{q}$	$q = q_0 + \mathrm{i}q_1 + \mathrm{j}q_2 + \mathrm{k}q_3$
$|q|$	norm of a quaternion q	$|q| = \sqrt{q_0^2 + q_1^2 + q_2^2 + q_3^2}$
$\bar{q}$	conjugate of a quaternion q	$\bar{q} = [q_0, -\mathbf{q}]$
$q \cdot r$ or qr	non-commutative quaternion multiplication of quaterions q and r (in general)	
$\hat{\mathbf{x}}$	unit vector of vector $\mathbf{x}$	$\hat{\mathbf{x}} \stackrel{!}{=} \frac{\mathbf{x}}{\|\mathbf{x}\|}$
$\mathbf{x}^{\mathrm{T}}, \mathbf{X}^{\mathrm{T}}$	transpose of vector $\mathbf{x}$ or matrix $\mathbf{X}$	$[\mathbf{X}^{\mathrm{T}}]_{ij} \stackrel{!}{=} [\mathbf{X}]_{ji}$
$\stackrel{\frown}{=}$	identification, correspondence	

Constants

Sym.	Meaning	Value	Source
c_0	speed of light in vacuum	$299\,792\,458\,\frac{m}{s}$	[74, p. 637]
$f_{☾}$	polar Moon flattening	$1/581.9$	[81, p. 898]
g_0	standard acceleration due to gravity	$9.806\,65\,\frac{m}{s^2}$	[72, p. 364]
G	NEWTONian constant of gravitation	$6.67428(67) \cdot 10^{-11}\,\frac{m^3}{kg \cdot s^2}$	[74, pp. 686–689]
k	GAUSSian gravitational constant	$0.017\,202\,098\,95$	[71, p. 58]
$r_{☾}$	volumetric mean Moon radius (MMR)	$1.73715 \cdot 10^6\,(\pm 10)\,m$	[81, p. 898]
$r_{☾\,pol}$	polar Moon radius	$1.73566 \cdot 10^6\,m$	[81, p. 898]
$r_{☾\,eq}$	mean equatorial Moon radius	$1.73864 \cdot 10^6\,m$	[81, p. 898]
$\mu_{☾}$	Moon's standard gravitational parameter	$4.902\,801\,076 \cdot 10^{12}\,(\pm 8.1 \cdot 10^4)\,\frac{m^3}{s^2}$	[78, p. 305]
$\mu_{☉}$	Sun's standard gravitational parameter	$1.327\,124\,400\,18 \cdot 10^{20}\,(\pm 8 \cdot 10^9)\,\frac{m^3}{s^2}$	[75]
τ_A	light time for 1 AU	$499.004\,783\,806\,(\pm 0.000\,000\,01)\,s$	[75]

Conversion Factors

Description	Conversion	Source
Astronomical Units $\longrightarrow$ Meters	$1\,AU = 1.495\,978\,706\,91 \cdot 10^{11}\,(\pm 3)\,m$	[75]
Julian Days $\longrightarrow$ Seconds	$1\,d = 86\,400\,s$	[75]
Julian Years $\longrightarrow$ Days	$1\,a = 365.25\,d$	[75]
Degrees $\longrightarrow$ Radians	$1° = 1° \cdot \frac{\pi}{180°}\,rad \approx 0.017\,453\,293\,rad$	

Symbols

Only symbols used in a global scope or context are listed here. Locally used symbols, i.e. within derivations and explanations, are explained where they are introduced in the continuous text.

Sym.	Meaning	Unit (in general)	Remark
a	semi-major axis of an orbit (KEPLERian element)	m	
$\mathbf{c}_{\text{direction}}$	camera direction vector (POV-Ray)	m	
$\mathbf{c}_{\text{pos}}$	camera position	m	
$\mathbf{c}_{\text{right}}$	camera-right vector (POV-Ray)	m	
$\mathbf{c}_{\text{up}}$	camera-up vector (POV-Ray)	m	
e	eccentricity of an orbit (KEPLERian element)	1	
$\mathbf{e}$	eccentricity vector		
$\mathbf{e}$	standard orthonormal base of $\mathbb{R}^3$		
$E(t)$	eccentric anomaly at epoch t (in the context of KEPLERian elements)	rad	
$\mathbf{h}$	orbital momentum vector	$\frac{\text{m}^2}{\text{s}^2}$	$\mathbf{h} = \mathbf{r} \times \dot{\mathbf{r}}$
i	inclination of an orbit (KEPLERian element)	rad	
M	mass (in general)	kg	
M_0	mean anomaly at initial epoch t_0 (KEPLERian element)	rad	$M_0 = M(t_0)$
$M(t)$	mean anomaly at epoch t (KEPLERian element)	rad	
$p(x, y, z)$	conversion function from rectangular coordinates $(x, y, z)^{\text{T}}$ to selenographic coordinates (ϑ, φ)	rad	$p \colon \mathbb{R}^3 \to \mathbb{R}^2, (x, y, z)^{\text{T}} \mapsto (\vartheta(x, y, z), \varphi(x, y, z))$
$\mathbf{p}$	a point within the ME/PA reference frame ($\mathbb{R}^3$) in general	m	
$\mathbf{p}(\vartheta, \varphi)$	conversion function from selenographic coordinates (ϑ, φ) to rectangular coordinates	m	$\mathbf{p} \colon \mathbb{R}^2 \to \mathbb{R}^3, (\vartheta, \varphi) \mapsto (x(\vartheta, \varphi), y(\vartheta, \varphi), z(\vartheta))^{\text{T}}$
$\mathbf{p}_\odot$	Sun position in the ME/PA reference frame	m	
q	quaternion (in general)	1	
$q_R(\alpha, \mathbf{u})$	rotation quaternion		see chapter 7
$\mathbf{r}(t)$	position vector at time t within the ME/PA reference frame (part of a state vector)	m	

$\dot{\mathbf{r}}(t)$	velocity vector at time t within the ME/PA reference frame (part of a state vector)	$\frac{\text{m}}{\text{s}}$	
$\mathbf{R}_x, \mathbf{R}_y, \mathbf{R}_z$	3-dimensional rotation matrix for a rotation around the x-, y- or z-axis, respectively		
t	epoch (in the context of KEPLERian elements)	d	
ϑ	selenographic latitude	rad	
μ	standard gravitational parameter of a celestial body (in general)	$\frac{\text{m}^3}{\text{s}^2}$	$\mu = GM$
$\nu(t)$	true anomaly at epoch t (in the context of KEPLERian elements)	rad	
Ξ	local tangent plane		see chapter 9
φ	selenographic longitude	rad	
ω	argument of periapsis of an orbit (KEPLERian element)	rad	
Ω	image plane of the camera (resulting in a plane of position vectors in the ME/PA reference frame)	m	for definition see equation 8.13 at page 108
Ω	longitude of the ascending node of an orbit (abbr. LAN; KEPLERian element)	rad	

Chapter 1

Thesis Background and Scope

1.1 Future Challenges of Space Exploration Missions

The launch of *Sputnik 1*, the first artificial satellite to be placed in Earth's orbit, on October 4th, 1957, marked the beginning of the Space Age — humanity was able to reach space for the first time. During the past 55 years, great achievements have been made in the field of space exploration: Mankind has been able to leave the Earth, and human beings have set foot on the Moon. First installations in space with a permanent human crew, like the *Mir* (Russian: *Мир*; lit. *Peace* or *World*) and the *International Space Station* (ISS), have been set up. Spacecraft such as *Voyager 1* traveled more than 150 Astronomical Units all the way to the outer frontier of our solar system and the limits of interstellar space. When the international community realized — especially after the end of the Cold War — that future missions will require international cooperation, the world's space agencies began to face a new era of space exploration.

In 2004, U.S. President George W. Bush announced the *Vision for Space Exploration*, the U.S. space policy [96]. One objective among others was the extension of human presence beyond *low Earth orbit*[1] (LEO) in conjunction with a return to the Moon by the year 2020 and the preparation for human exploration of Mars. As one consequence of the Vision for Space Exploration, NASA established the *Constellation Program* (abbrv. CxP) in the same year. CxP contained an ambitious plan for a human mission to Mars, using the Moon as a testbed for

---

Chapter Image: Artist's concept of lunar lander *Altair* (NASA Constellation Program, cancelled). ©2007 NASA, License: Public Domain. Available at http://www.nasa.gov/mission_pages/constellation/altair/altair_concept_artwork.html.

[1] Low Earth orbit: A geocentric orbit with apogee and perigee below 2 000 km altitude above mean sea level.

future human and robotic missions [84, p. 3] (this kind of strategy is called the *Moon-first approach*²). In 2006, NASA announced plans to establish a permanent lunar outpost [100].

Six years later, in 2010, President Barack OBAMA replaced the Vision for Space Exploration with the new U.S. National Space Policy, canceling the Constellation Program on recommendation of the *Review of United States Human Space Flight Plans Committee* (also known as the *Augustine Commission*), which stated that the CxP is not feasible within the assigned budget. President OBAMA replaced the Moon-first approach with multiple destinations in the Solar System, the so-called *flexible-path approach*.

> *Humanity's interest in the heavens has been universal and enduring. Humans are driven to explore the unknown, discover new worlds, push the boundaries of our scientific and technical limits, and then push further. NASA is tasked with developing the capabilities that will support our country's long-term human space flight and exploration efforts. We have learned much in the last 50 years, having embarked on a steady progression of activities and milestones with both domestic and international partners to prepare us for more difficult challenges to come. Our operations have increased in complexity, and crewed space journeys have increased in duration. The focus of these efforts is toward expanding permanent human presence beyond low Earth orbit.*
>
> — 2011 NASA Strategic Plan [97, p. 7]

NASA's plans have been used as an example for the future development of space exploration missions at this place, since it is the foremost space agency with respect to the exploration of the solar system so far. However, in a long-term and broader view over all space agencies, aims of current and future missions will be a return to the Moon, the human and robotic exploration of near-Earth asteroids as well as Mars, and finally the extension of human presence in space by setting up habitats on the Moon and Mars, even if those plans are delayed due to financial or political amendments.

In 2004, the 14 leading space agencies formulated their common goals within a 25-year strategy paper [87] as a vision for peaceful robotic and human space exploration. Common international goals, bringing benefits to all humans, are outlined as follows [87, p. 2]:

- search for life
- extend human presence
- develop exploration technologies and capabilities
- perform science to support human exploration
- stimulate economic expansion

---

² The opposite approach, focusing on Mars as the first celestial body to be inhabited, is called the *Mars-first approach*.

- perform space, Earth and applied science
- engange the public in exploration
- enhance Earth safety

A short summary of key objectives and challenges on a per-target basis for the ambitions of future missions is given in the paper, too:

Target	Key Objectives	Challenges
Mars	Search for life. Advance understanding of planetary evolution. Learn to live on other planetary surfaces.	Significant technology advancements are essential for safe and affordable missions. Radiation risk and mitigation techniques must be better understood. Highly reliable space systems and infrastructure are needed. Demonstrated ability to use local resources is essential.
Moon	Characterize availability of water and other resources. Test technologies and capabilities for human space exploration. Advance understanding of solar system evolution. Utilize the Moon's unique importance to engage the public.	Expenses associated with extended surface activities.
Near-Earth Asteroid	Demonstrate innovative deep space exploration technologies and capabilities. Advance understanding of these primitive bodies in solar system evolution and the origin of life. Test methods to defend the Earth from risk of collisions with near-Earth asteroids.	Need to better understand and characterize the asteroid population. Technology advancements are needed before missions to asteroids.

LaGrange Points/Cis-Lunar Space	Expand capability of humans to operate in this strategic region beyond low-Earth orbit.	Understanding the benefit of human presence vs. robots.
	Demonstrate innovative deep-space exploration technologies and capabilities.	

Table 1.1 This table shows common future targets of space exploration missions including their key objectives and challenges. ("Summary of the Destination Assessment Activity", quoted from [87, p. 15])

1.2 Necessity of New Navigation Technologies

In our pursuit of the aforementioned goals, the development of advanced technologies for future space exploration missions is inevitable, including development of technologies for the establishment and the provision of outposts, as well as the exploration of locally limited phenomena on moons, planets and asteroids. One requirement for these missions is a system for autonomous, precise and secure navigation and landing on celestial bodies, for example, "[...] to enable targeted landing of payloads near to each other and in-situ resources" [88, p. 1].

The European Space Agency (ESA) takes the automatic precision navigation and landing into account in their *Aurora* program as one of the "technologies Europe must have" [86, p. 6]. The German national space program [82, pp. 12–14] adopted advanced technologies for future space exploration missions as national science goals.

Today's spacecraft navigation systems rely on measurements and calculations done on Earth. Navigating a spacecraft in space involves two essential processes: First, a determination and prediction of the spacecraft position and velocity (state vector), which is called orbit determination, has to be done. The subsequent step is the flight path control, altering the spacecraft velocity (and by association its trajectory).

The step of orbit determination comprises the finding of the actual and accurate spacecraft's orbital elements as well as the determination and accounting for perturbations in its natural orbit. By comparison of the predicted spacecraft's trajectory (based on the orbit determination) and the destination trajectory, necessary velocity changes can be calculated and commanded to the spacecraft (flight path control).

Contemporary orbit determination involves several measurements (e.g. Doppler and signal latency measurements, Very Long Baseline Interferometry, precision ranging, etc.) done from

Earth[3]. Since Earth's orbital elements and motions are well known, every state referring to Earth can be translated to other reference frames. An astrodynamical model, which will be refined with each single measurement, is built and maintained on Earth. Based on this model, the necessary trajectory predictions can be made. The big disadvantage of this technique is the total dependency on Earth.

Hence, one critical part of future space exploration technologies will be new navigation systems for an accurate in-situ position determination of spacecraft in the orbit or on the surface of celestial bodies, as well as for a precise and safe landing procedure. It is important that these measurements can be done completely independent of Earth, enabling missions without permanent contact to Earth or human intervention, as existing systems rely on the tracking and flight path control from Earth. This ability is one key technology for a safe and autonomous navigation and landing of robotic — or perhaps manned — spacecraft.

A promising approach involves optical navigation technologies, because their measurements are completely independent from Earth, cutting away long signal latencies and the aforementioned constraints. Furthermore, such an optical navigation system is not affected by volatile communication and tracking conditions (i.e. caused by occultation effects, transits, electromagnetic storms and so on). In addition, pictures taken by the optical navigation system can be used for the selection of a safe landing place. These advantages are currently bringing the development of optical landing techniques to center stage worldwide.

1.3 Ambitions of the German Aerospace Center (DLR)

In the future plans of the German Aerospace Center (German: *Deutsches Zentrum für Luft- und Raumfahrt*, abbrv. DLR), the Moon is a primary target for the German space program in the coming years. DLR established the interinstitutional project *Autonomous Terrain-based Optical Navigation* (ATON) several years ago, aiming to develop a holistic navigation solution for the three stages of a lunar landing procedure (described in the next paragraph). The DLR Institute of Space Systems in Bremen hosts the *Testbed for Robotic Optical Navigation* (TRON), a laboratory facility for the generation of real-time image data of the Moon's surface.

Assuming that the spacecraft is in a lunar parking orbit, a lunar landing procedure can be divided into three stages[4]: At first, the landing procedure begins with a Descent Orbit Injection maneuver (DOI), the short firing of the engines to initiate the following unpowered descent from parking orbit to lunar surface, until the spacecraft reaches an altitude of 10 to 15 km

---

[3] A good overview in layman's terms of the measurements done for orbit determination can be found at http://www2.jpl.nasa.gov/basics/bsf13-1.php; this provides the basis for the contents in this section.

[4] The overview of the stages of a lunar landing has been provided by Bolko Maass (DLR); this explanation is a rough translation of his explanation into English.

($\Delta v \approx 20\ldots30\,\frac{m}{s}$)[5]. For a precise landing, a position determination with an accuracy of about 100 m is necessary, and this cannot be achieved by conventional technology. One approach is the navigation based on landmarks visible on images of the surface, while the spacecraft maintains a catalog of references aboard.

In the second phase, the spacecraft's relative velocity in relation to the Moon's surface will be reduced and the spacecraft will be located about 1 000 to 1 500 m over the target destination; this braking action is called a *Powered Descent Initiate* (PDI). The engines are active during the entire phase ($\Delta v \approx 1.7\ldots1.8\,\frac{km}{s}$). During this phase a navigational solution for the descent path is needed with an accuracy about 0.5 to 1 % of the current altitude. A solution could be the *terrain-relative navigation* (TRN) based on the tracing of a significant, but unknown, point on the Moon's surface. A high frequency (>1 Hz) estimation of the spacecraft's altitude and velocity can be derived in conjunction with an acceleration sensor.

The last stage consists of the selection of a secure landing place within the given radius of precision (so-called *Hazard Avoidance*, abbrv. HA) and the landing itself. The navigation system maintains a map of the target area and compares the map with the optical input. All errors arising from the preceding stages can be corrected during the ongoing descent. An in-situ selection of a secure landing place is inevitable due to the lack of maps with sufficient detail, which would enable an a priori selection of a landing place — even in the conceivable future. Considerations regarding, for example, the absence of surface irregularities and stones as well as a sun-illuminated landing place, have to be given to ensure a safe in-situ selection of a landing place, as well as a rating of specific fuel consumption for maneuvers to alternative landing sites. During this stage, the optical resolution will increase during descent, while maneuverability will decrease.

Within the DLR ATON project, new navigation technologies for a lunar landing procedure shall be developed with distinct approaches for the several stages (i.e. landmark detection, feature tracking, stereo imaging, sensor technology). ATON's principal objective is to develop an image processing and navigation software for estimating the position and orientation of a spacecraft, as well as to demonstrate a prototype in a laboratory environment. This environment is called the *Testbed for Robotic Optical Navigation* (TRON) [90].

TRON is a simulation chamber allowing the simulation of geometric, optical and astrodynamical conditions during a landing procedure, starting with the approach to the celestial body and ending with the landing itself, for a myriad of celestial bodies, including the Moon, Mars and asteroids. The testbed is an isolated lab, equipped with a six degrees of freedom (DOF) industrial robot on a rail system and a 5 DOF illumination system. The lab can be equipped

---

[5] Delta-v (Δv) is a widely used measure for the capability of a spacecraft to make an orbital maneuver (to change from one trajectory to another). Δv is mass-invariant and independent from technical details of the propulsion system. Hence, a declaration of spacecraft mass, thrust or fuel level is unnecessary.

1.3 Ambitions of the German Aerospace Center (DLR)

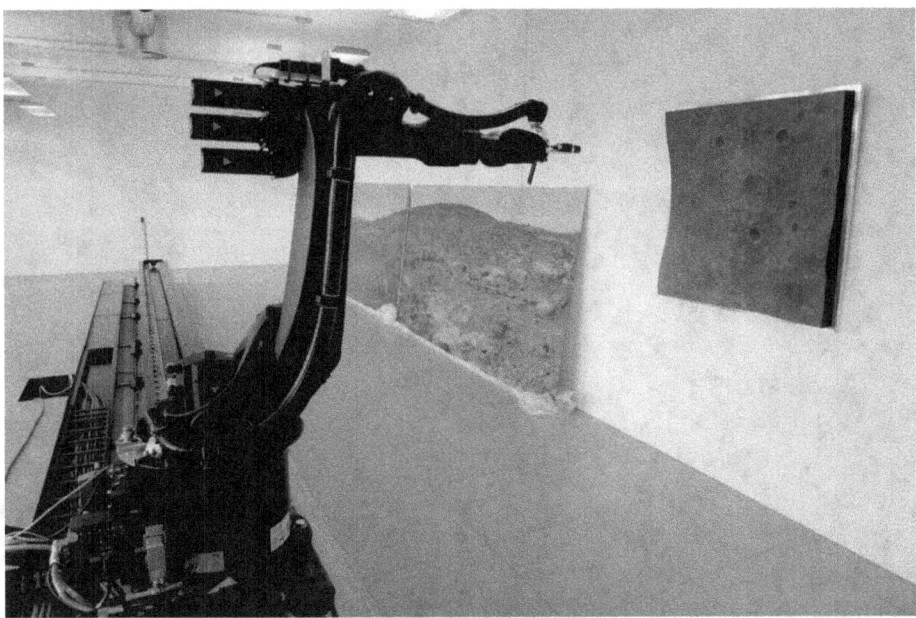

(a) Early stage: 6 DOF industrial robot on a rail system and one wall-mounted, preliminary 3D surface tile

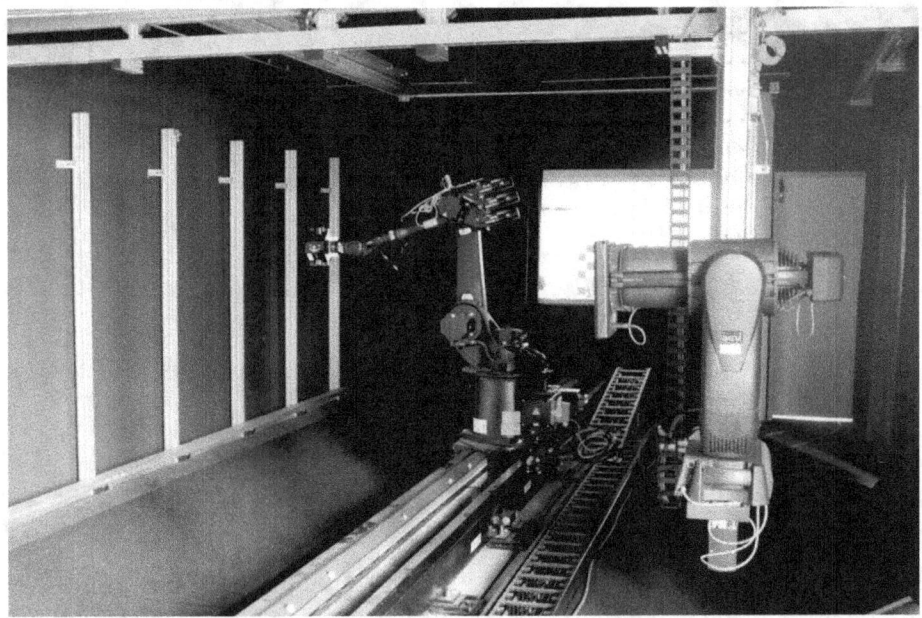

(b) Advanced stage: 6 DOF industrial robot (left) and the 5 DOF illumination system (right) with installed bearing structure for the surface tiles. All walls have now been painted black to reduce diffuse reflection.

Figure 1.1 TRON facility in an early and advanced stage of construction. © DLR; pictures reproduced with friendly permission.

1 Thesis Background and Scope

Figure 1.2 A sample 3D surface tile of the Moon used in an advanced construction stage of TRON, illuminated with the 5 DOF illumination system: Realistic shadows are cast. © DLR; picture reproduced with friendly permission.

with wall-mounted 3D surface tiles (terrain models) of celestial bodies, which are made by milling digital elevation models (DEM) out of foam material. This way, TRON can be used for all celestial bodies, for which sufficient DEMs exist and 3D surface tiles have been fabricated. A schematic diagram of TRON can be seen in figure 1.3.

Using this configuration, real-time high-quality shadows can be simulated and tests with 3D sensors (i.e. stereo cameras, LIDAR sensors) are possible. Such sensors are mounted on the tool center point of the robot, allowing them to be flown over the terrain in the fashion of a virtual spacecraft. In conjunction with dSPACE real-time simulation hardware, hardware-in-the-loop simulations, qualifying and validating sensors and the later navigational solution, can be done for all stages of a landing procedure.

This project is one branch of the ongoing research of the DLR towards a new optical navigation system by landmark detection and matching with a spacecraft-maintained surface feature catalog. In this context, craters are favored geographical features on solid planetary surfaces. It can be assumed that they exhibit very characteristic shapes and that they are stable over

1.3 Ambitions of the German Aerospace Center (DLR)

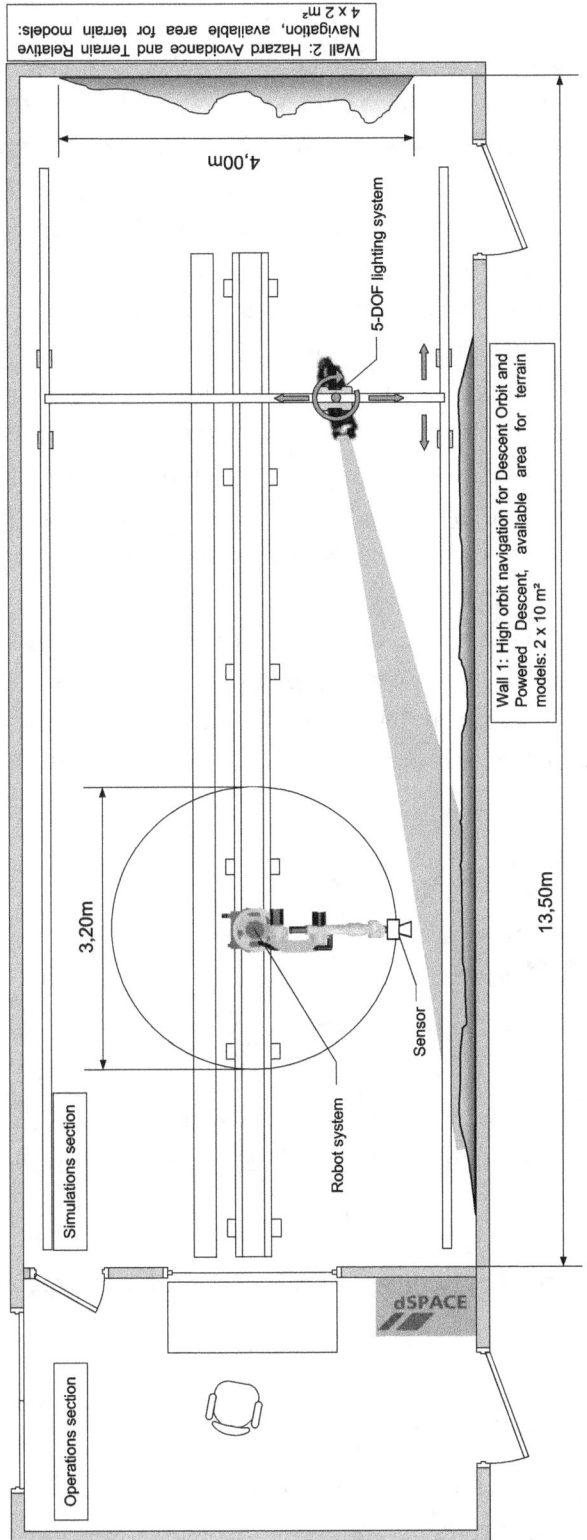

Figure 1.3 Schematic overview of the *Testbed for Robotic Optical Navigation* (TRON) laboratory, located at the DLR Institute of Space Systems in Bremen, as seen from above. The lab is divided into two sections: Operations (the simulation control room) and simulations section. At the bottom and the right border of the simulation section's illustration the wall-mounted surfaces tiles are depicted. These tiles can be illuminated by a 5 DOF illumination system, as implied by the red beam. The scene is then captured with sensors, i.e. optical or LIDAR sensors, which are mounted at the tool center point of the 6 DOF industrial robot on a rail system. Reproduced with friendly permission from [90].

a long duration in terms of shape, structure and positioning, as long as effects attributable to erosion, physical and chemical weathering and denudation are absent or negligible on the celestial surface. Hence, they can be used as landmarks in a navigational context.

Before captured craters can be compared against a catalog, they have to be perceived by a software out of images taken by the optical instruments. As MAASS et al. [90] elucidate, existing algorithms require the satisfaction of some essential conditions, for example, a uniform illumination over the entire image or a camera alignment perpendicular to the surface. Additionally, they are too resource-intensive for current space-qualified hardware, resulting in failed timing requirements needed for navigation purposes.

Furthermore, modern crater detection algorithms rely on some kind of edge detection of the crater rim for crater identification. This seems a logical strategy, since most of the crater rims on planetary surfaces can be perceived as circles or ellipses (according to the camera angle in relation to the surface plane and impact angle of the crater's cause). However, this model is too restrictive to detect all craters; actually only a small subset of all craters are detectable with such an approach, because not all crater rims are well-delimited (e.g. due to size, smoothness, non-circular rims, stage of erosion or disturbances caused by other surface features).

With these strict limitations in mind, the DLR is developing a new edge-free, scale-, pose- and illumination-invariant crater detection algorithm [90]. This algorithm relies on the detection of crater-characteristic contrast areas due to a non-perpendicular illumination angle as illustrated in figure 1.4.

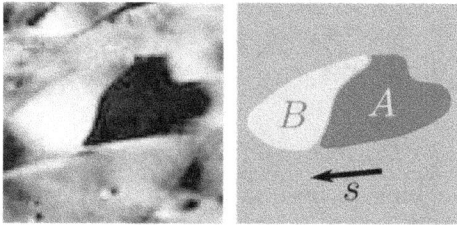

Figure 1.4 Extraction of crater contrast areas out of a sample picture; the vector shown beyond is the local solar illumination vector. © DLR; reproduced from [90, p. 605] with friendly permission.

This new algorithm shall remove the limitations of existing algorithms; moreover it is believed to

- increase the detection rate of craters,
- provide a performance enhancement,
- offer the possibility of a FPGA (*Field Programmable Gate Array*) implementation and
- lower the crater-diameter detectability threshold.

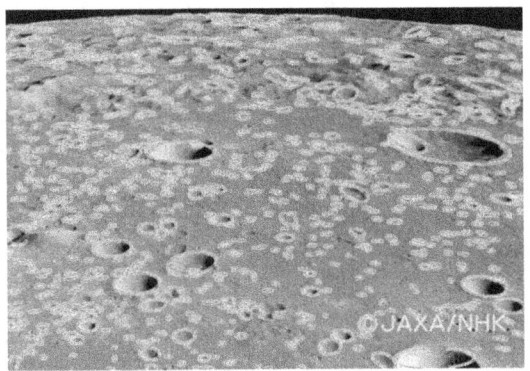 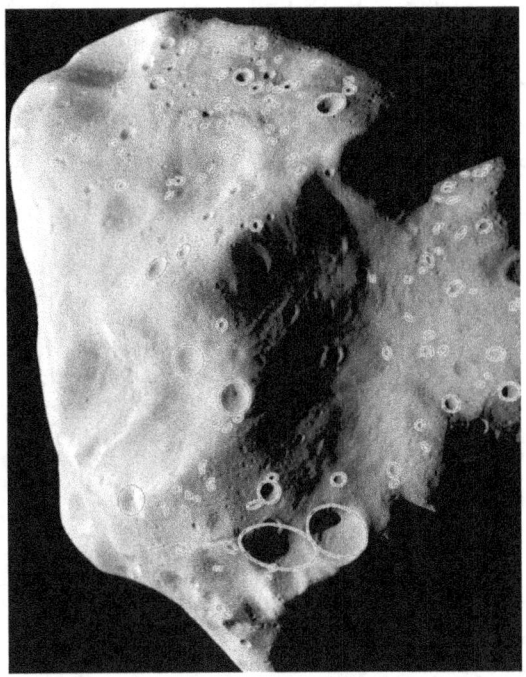

(a) The Moon's surface (Base Image: JAXA/Kaguya/SELENE)

(b) Asteroid Letetia (Base Image: ESA/Rosetta Mission)

Figure 1.5 First results of the new crater detection algorithm on pictures of real celestial surfaces. © DLR; pictures reproduced from [90, p. 608] with friendly permission.

Preliminary studies show promising results with respect to the previously mentioned goals, since the algorithm's performance is one hundred times higher compared to existing algorithms and the crater diameter detection threshold could be lowered to only 6 pixels (current thresholds are about 10 to 20 pixels) [90, p. 605]. First results on pictures of real scenarios can be seen in figure 1.5.

Further work has to be done regarding systematic tests of the detection rate, stability, performance and details of the algorithm, as it is currently in an early development stage. To promote further development, the possibility of generating realistic surface images with well-known illumination conditions is essential, as well as a technique for estimating the illumination direction on these pictures. TRON will be an indispensable tool once it is completed. However, it has some limitations, for example, the simulation of non-uniform illumination conditions caused by the surface curvature or the curvature itself are not possible. In addition, TRON's surface coverage is limited, both in terms of its conception and its ability to switch to other celestial bodies, as all wall-mounted surface tiles have to be changed.

One solution would be a simulation software; nevertheless such software gives rise to new limitations: Currently, required digital elevation models (DEM) are only available in sufficient resolutions for the Moon and Mars. More importantly, no sensor tests or qualifications are possible solely based on software. This means that only a combination of the two approaches can overcome the limitations; their results are directly comparable.

1.4 Thesis' Contribution to Novel Navigation Systems

Optical navigation technologies based on images taken by cameras represent only one category of terrain-relative navigation (TRN) methods, the passive-sensing TRN. In general, TRN is thought to provide estimations of navigational parameters (position, velocity, attitude) by comparing measurements with known surface features of a celestial body. Hence, a priori reference maps are needed for TRN, which can differ, depending on the chosen TRN approach. The second category is active-sensing TRN approaches, which rely on active-ranging techniques (LIDAR or radar altimetry). [89, p.1]

Both categories have advantages and disadvantages: Cameras, which are needed as sensors in the passive imaging, are technologically mature and put small demands on the spacecraft's engineering regarding power consumption, mass and volume; measurements are possible from all altitudes for most of the methods. However, they impose an operation at solar illumination, since they cannot operate in the dark. Active ranging, in contrast, eliminates this constraint, but it is more resource-consuming than passive sensing and it imposes limits on the applicable operating range, resulting in limitations to the altitude, in which the sensors are operational. [89, p.2]

The aforementioned DLR research aims at the development of a TRN system based on landmark detection, which is basically an advanced crater-pattern matching approach to TRN. Nevertheless, there are many other approaches to TRN besides crater-pattern matching, which require different inputs and produce distinct outputs. An overview of all current active- and passive-sensing TRN approaches is given in figures 1.6 and 1.7, respectively.

Common to all passive-sensing TRN approaches is that all inputs result out of camera images. To date, no applicable software has been developed for the generation of artificial renderings of realistically illuminated planetary surfaces while determining the illumination direction. Also, as TRON is not finished yet, appropriate test images of planetary surfaces with well-known illumination directions for algorithms tests and error estimations are rare.

Thus, the primary objective of this thesis is the development of a surface illumination simulation software for solid planetary surfaces. This simulation software is a mandatory and essential first step enroute to the generation of realistic test imagery for all passive-sensing TRN approaches. Furthermore, this software will enable the DLR to trial their new crater detection

1.4 Thesis' Contribution to Novel Navigation Systems

Sensing Modality	Active-Sensing			
Objective	Position Estimation			Velocity Estimation
Sensor Type	Imaging LIDAR		Altimeter	Imaging LIDAR
Approach	Shape-Signature Pattern Matching	Range-Image to DEM Correlation	Altimeter-to-DEM Correlation	Consecutive Range-Image Correlation
Inputs	- Range Image - Motion Correction Data - Shape-Signature Data (Based on 3D Map)	- Range Image or Scans - Motion Correction Data - Absolute Attitude Estimate - Digital Elevation Map	- Altimetry Swath - Motion Correction Data - Absolute Attitude Estimate - Digital Elevation Map	- 2 Range Images - Motion Correction Data - 2 Attitudes
Output Estimates	Absolute Position and Attitude	Absolute Position	Absolute Position	Average Horizontal and Vertical Velocity
Strengths	- general approach solves for position and attitude w/o prior knowledge of these measurements - independent of lighting conditions	- independent of lighting conditions - more robust than Altimeter-to-DEM Correlation	- independent of lighting conditions - sensors likely to work at higher altitudes (possibly up to 100 km)	- independent of lighting conditions
Limitations	- long processing time - more general than needed - significant terrain relief required - LIDAR less mature than camera	- scanner, gimbal or imaging array required - LIDAR less mature than camera	- long contour required - LIDAR less mature than camera	- image overlap required
Non-Space Application	Object Recognition from Range Data		Raytheon Cruise Missile TERCOM	

Figure 1.6 Overview of active-sensing approaches for TRN. Only shown in completion to figure 1.7, since these methods rely on ranging techniques, which are not applicable to the thesis software. (Image based on the table in [89, p. 4])

algorithm in conjunction with various illumination direction estimation techniques on the produced pictures. That is why a further thesis goal is to give an overview of existing approaches for the determination of the illumination direction with some first algorithm implementations and tests. All work has been done in the context of using the Moon as a representative of the class of solid planetary surfaces with a significant number of craters.

The primary objective comprises the sighting, conditioning and selection of available lunar topography data, as well as a conversion of these data sets for the creation of a virtual, 3D surface model of the Moon. This part involves considerations of suitable coordinate systems and transformations. The NASA/NAIF SPICE toolkit will be used for the geometrical and astrodynamical calculations.

While the first objective conduces image synthesis, the second objective represents the image analysis — and therefore an approach to illumination direction estimation. This part of the thesis shall give a brief overview of this field of research, whereby the available literature shall be primarily cited. Optionally, a selection or customization/development of methods and algo-

1 Thesis Background and Scope

Sensing Modality	Passive-Sensing					
Objective	Position Estimation				Velocity Estimation	Velocity and Attitude-Rate Estimation
Sensor Type	Camera				Camera	Camera
Approach	Crater Pattern Matching	Scale-Invariant Feature Transform (SIFT) Pattern Matching	Onboard Image Reconstruction for Optical Navigation (OBIRON) Surface-Patch Correlation	Image-to-Map Correlation	Descent Image Motion Estimation Subsystem (DIMES) Consecutive Image Correlation	Structure from Motion Consecutive Image Correlation
Inputs	- Descent Image - Crater Landmark Database	- Descent Image - SIFT Landmark Database	- Multiple Overlapping Orbital Images - Descent Image - Lander Attitude - Lander Altitude	- Map Image - Descent Image - Lander Attitude - Lander Altitude	- 3 Descent Images - 3 Attitude Estimates - 3 Altitude Estimates	- 2 Descent Images - 2 Attitudes
Output Estimates	Absolute Position and Attitude	Absolute Position and Attitude	Absolute Position and Attitude Update	Absolute Horizontal Position	Average Horizontal Velocity	Average Velocity and Angular Rate between Images
Strengths	- insensitive to illumination changes - no attitude or altitude measurements required	- general representation should work for all terrains including ones without craters - no attitude or altitude measurements required	- general representation should work for all terrains including ones w/o craters - built-in accomodation of illumination changes and terrain relief	- general representation should work for all terrains including ones w/o craters - requires just one orbital image and no 3D modeling or rendering	- general representation should work for all terrains	- no attitude estimation required - general representation should work for all terrains - fast implementation and very accurate
Limitations	- solar illumination required - cratered terrain required	- solar illumination required - illumination changes between image and map not well tolerated - large out-of-plane rotations degrade performance	- solar illumination required - multiple overlapping images of landing site required - rendering of landing site required prior to landing - attitude/altitude estimations required	- solar illumination required - possibly sensitive to large illumination changes and terrain relief	- solar illumination required - overlap between consecutive images required	- solar illumination required - overlap between consecutive images required
Non-Space Application				Terrestrial Rover Navigation	Raytheon Cruise Missile DSMACS	

Figure 1.7 Overview of passive-sensing approaches for terrain-relative navigation (TRN). The thesis software can produce renderings of planetary surfaces for all kinds of passive-sensing approaches for TRN, indicated by fields with yellow and orange backgrounds. The crater detection algorithm currently developed by DLR within the ATON/TRON project uses the Crater Pattern Matching approach, which is marked in orange. As it can be inferred from this overview, the thesis software can be used not only for the current DLR project, but for all other TRN approaches. (Image based on the table in [89, p. 4])

rithms shall be done; first algorithms shall be implemented, tested and rated with the images resulting from the new surface-illumination simulation software.

With the new software resulting out of this thesis work, a basis will be established for the generation of appropriate renderings of planetary surfaces at changing illumination conditions, which can be used for all passive-sensing TRN approaches. This software will be developed to produce renderings not only for the Moon, but for all celestial bodies, for which digital elevation models exist. This way, the software will be a worthwhile, reusable and flexible tool for the research of TRN technologies. See figure 1.7 for a visualization of all TRN approaches for which the new software is applicable. The DLR approach is highlighted in orange in this figure, while all other approaches for which the software is additionally applicable are highlighted in yellow. It can be inferred that the new software has far-reaching implications for use beyond the thesis' scope.

All elaborations are considered to be *proof of concept*, so no considerations have been given to computing time/complexity, resource utilization or algorithm optimality, and this is intentional.

Introducing the Moon Surface Illumination Simulation Framework (MSISF)

2.1 General Concept

The *Moon Surface Illumination Simulation Framework* (MSISF) has been developed in pursuing the primary objective of this thesis, the development of an illumination simulation software for the Moon's surface. Initially, the development of a more general software for the illumination simulation of solid planetary surfaces was envisaged. This software should use 3D modeling techniques to create surfaces of artificial celestial bodies with a significant number of craters, which would later be illuminated with 3D rendering and ray tracing techniques. Since for some time highly precise LIDAR[1] data of lunar topography has been available, the Moon has been chosen as one representative of this class of celestial bodies. This way, a separate development

---

Chapter Image: *Another way of gathering data regarding the Moon's shape.* Original image description: Wallops automatic programmer being monitored by F.H. Forbes, July 29, 1950. Doppler radar recorders are behind Forbes. Photograph published in A New Dimension; Wallops Island Flight Test Range: The First Fifteen Years by Joseph Adams Shortal. A NASA publication (page 100). ©1950 NASA, License: Public Domain. Available at http://lisar.larc.nasa.gov/UTILS/info.cgi?id=EL-2002-00247.

[1] *Light Detection and Ranging* is a technique for measuring a distance to a target using laser beams. For this purpose, laser rays are shot to a target; the light will be usually reflected by the physical principle of backscattering. Measuring the latency between outgoing and incoming light, the distance between target and observer can be derived using the known speed of light in a vacuum.

of an artificial 3D surface generator was no longer required.

The development of the MSISF took place from a practical point of view: As much as possible, already existing software components should be used for this software to ensure a rapid development, fitting in the narrow timetable of a master's thesis. In fact, nowadays a real myriad of 3D modeling, rendering and ray tracing tools exist. However, there are a lot of incompatibilities, restrictions and performance issues connected with the tight integration of these tools into a distinct software development process. Software architecture planning has to be well conceived to ensure a properly functioning collaboration of the individual software components.

2.2 Development Milestones

Out of these considerations, the software to be developed is not only a single piece of binary code, but rather a *framework* of components for distinct tasks. Consequently, the developed software was dubbed *Moon Surface Illumination Simulation Framework* (MSISF). Figure 2.1 shows an overview of the layout and functional interaction of the MSISF components. The MSISF mainly consists of the *Moon Surface Illumination Simulator* (MSIS), which is the main user interface. MSIS is a Windows command-line application, which makes use of the Microsoft .NET Framework 4.0 and has been written in C#. This application invokes all steps to produce the later renderings of the Moon's surface, which involves, among other things, the astrodynamical calculus, time conversion routines, the selection algorithm for the 3D mesh data used and the invocation of POV-Ray as 3D rendering engine as well as methods for obtaining and displaying meta information about the rendering.

MSIS compiles the used 3D data out of its own data repository, which is named *pattern repository*. This pattern repository is built out of so-called *digital elevation models* (DEMs) from external sources with the help of the PHP scripts `ldem_mysql_insert.php` and `generate_pattern.php` within the MSISF pattern preparation process. Additionally, the MSISF consists of the NASA NAIF SPICE toolkit, used for the astrodynamical calculus as well as several supporting MATLAB scripts, which were used for testing and rendering preparation during this thesis work.

The MSISF development has been divided into six pieces of theoretical and practical work as subsequent steps towards the final software framework. These steps will be described in brief in the following paragraphs.

2.2.1 Reference Frame and Selenographic Coordinates

Before the start of any work regarding a software development, preliminary considerations of the theoretical foundations of the work to be done have to be made (chapter 3). First, the

2.2 Development Milestones

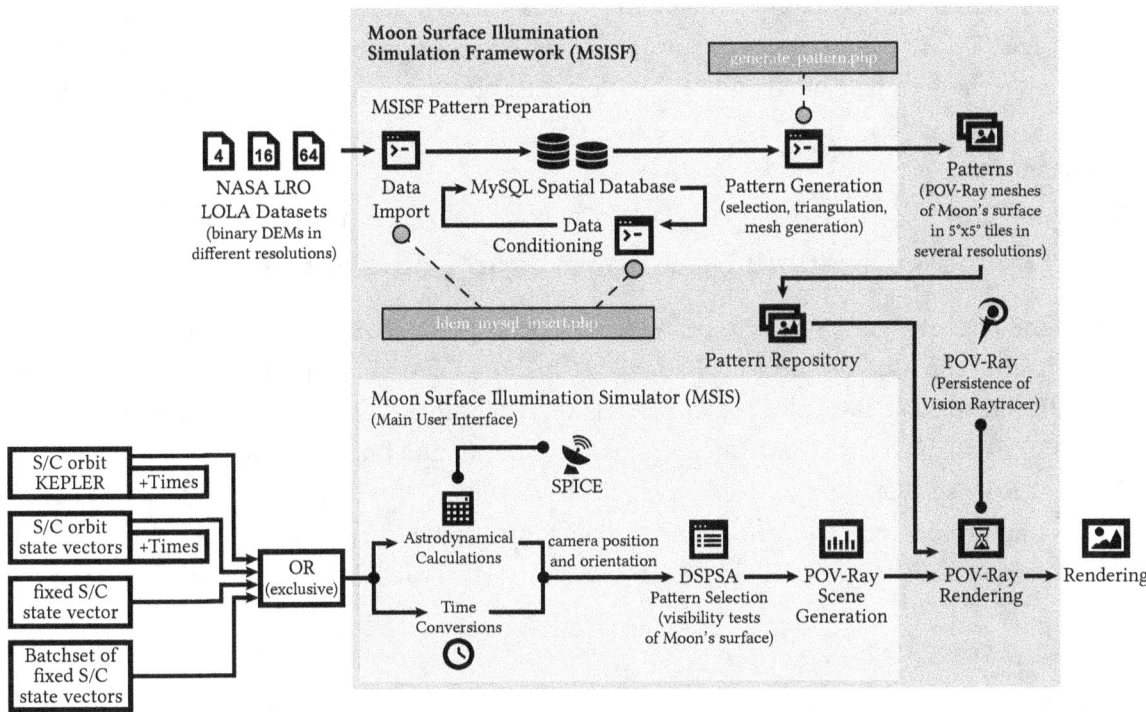

Figure 2.1 Overview of the components of the *Moon Surface Illumination Simulation Framework* (MSISF).

predetermination of the coordinate system to be used for referencing positions in space is essential for a coherent and standard-conform usage of all spatial coordinates. The author will use the so-called *Mean Earth/Polar Axis* (ME/PA) reference frame, which is the reference frame for the Moon recommended by the International Astronomical Union (IAU). Section 3.1 will give an introduction to this reference frame.

Out of the definitions of this reference frame, an additional coordinate system for the easy referencing of surface points can be derived, which is similiar to the geographic coordinate sytem used on Earth by defining two numbers as latitude and longitude. This coordinate system for the Moon is called the *selenographic coordinate system*; latitude and longitude value pairs for the Moon's surface are simply called *selenographic coordinates*. The conversion from spatial coordinates within the ME/PA reference frame to selenographic coordinates and vice versa is explained in section 3.2.

2.2.2 Lunar Topography Data Acquisition

First of all — after the decision was made to use lunar LIDAR data — appropriate digital elevation models (DEM)[2] for the Moon's surface have to be researched and examined, and will be discussed in the first part of chapter 4. Without meaning to anticipate this chapter, the lunar digital elevation models (LDEMs) of the NASA Lunar Reconnaissance Orbiter (LRO) mission have been used.

2.2.3 Database Creation, Data Import and Conditioning

The second part of chapter 4 discusses how to harness the NASA LRO LOLA data to generate a 3D model of the Moon's surface. To achieve this, the DEMs have to be read and converted to a utilizable format. Basically, the NASA LDEMs contain a point cloud, which is referenced against an ideal sphere with a defined radius. This point cloud needs to be sorted; some parts of it need to be selected in a simple and efficient way. It was obvious to use a spatial database, which is optimized for such tasks. The decision was made to use MySQL for this purpose; the second part of chapter 4 elucidates all steps towards building a global lunar topographic database.

2.2.4 Surface Pattern Generation

3D tiles of the lunar surface, so-called *surface patterns*, will be made out of the digital elevation data stored in the spatial database, using the 2D DELAUNAY triangulation and producing surface patterns in the form of *triangulated irregular networks* (TIN) for the entire lunar surface. These

---

[2] At the moment, no scientific notation standard has emerged for the usage of the terms *digital elevation model* (DEM), *digital terrain model* (DTM) or *digital surface model* (DSM). The term *digital elevation model* (DEM) is often used as a collective term for both the *digital terrain model* (DTM) and the *digital surface model* (DSM). The author subscribes to the common opinion that DTMs are data sets without capturing lifeforms (trees, plants etc.) and artificial structures (buildings, streets and so on) on a surface and DSMs — in contrast — are data sets captured with all objects on the surface (cf. figure 2.2).

Figure 2.2 Visualization of the difference between a digital terrain model (DTM) and a digital surface model (DSM).

Since currently very few artificial structures exist on the Moon, all DSMs and DTMs will be virtually the same for the Moon, so this is rather a question of notation than contentual difference. Henceforth, the author will use the general term digital elevation model (DEM).

surface patterns are stored in a file format which can be parsed by POV-Ray, an open source 3D ray tracing and rendering software (chapter 5).

After all surface patterns for the entire lunar surface have been generated and stored[3], the *Moon Surface Illumination Simulator* (MSIS), which is the core software component of the MSISF, does all the necessary calculations (astrodynamical calculus, orbit and state determination, time conversions), selects all the required surface patterns (based on what will be visible by a virtual spacecraft/camera in the orbit around the Moon) and initiates the rendering process by invoking POV-Ray (chapter 6).

2.2.5 Surface Pattern Selection, Assembling and Rendering

As said before, the renderings of the illuminated lunar surface will be done by using a virtual spacecraft/camera orbiting the Moon. Based on its orbit, a given time and some additional parameters for orientation and geometry of the camera, the information about which parts of the Moon's surface will be visible on the later image can be derived using ray tracing techniques. This is an indispensable step to cope with the enormous amount of LDEM data for higher resolutions, since only really required data has to be included in the rendering process. To determine the camera position and orientation, several predefinitions regarding a spacecraft orientation and rotation model have to be established (chapter 7).

2.2.6 Output of the Local Solar Illumination Angle

Additional considerations have been necessary regarding the dynamical selection of the required surface patterns, as well as for the determination of the local solar illumination angle. The technique for the selection of the required surface patterns is named *Dynamical Surface Pattern Selection Algorithm* (DSPSA) and is documented in chapter 8, while the method used for the determination of the local solar illumination angle is discussed in chapter 9.

2.2.7 Result Discussion

Finally, this thesis will end up with a result discussion, several suggestions for improvements of the MSISF and a comparison with latest results of the DLR research, especially with reference to the TRON facility, as well as latest achievements in the generation of digital elevation models using stereoscopic imaging techniques (chapter 10).

---

[3] This process has to be done only once per LDEM resolution. The MSISF delivered to the DLR contains a pre-processed surface pattern repository, so there is no need to re-generate the surface patterns, unless a new LDEM resolution is added, existing LDEM data has changed or surface patterns for another celestial body are generated.

2.3 Preparatory Remarks

2.3.1 Software and Programming Languages Used, Development Environment

In the course of this book, the following chapters will provide information about the performance of the used software components (e.g. rendering times, MySQL query execution performance profiles, etc.). In a world of rapidly changing IT systems, such information is only useful with corresponding information of the used development and test environment, since it has a significant impact on the overall performance. The database queries given in chapter 4, for example, could be accelerated by a couple of times by replacing the conventional harddrives, which are used as data storage for the MySQL database on the test machine, with solid-state disks (SSDs). At the time of writing this thesis, SSDs with sufficient storage capacity (> 250 GiB) have just became affordable in the consumer sector.

All software development and testing has been done on a single, standard personal computer using Microsoft Windows 7 x64 Professional as operating system; table 2.1 gives an overview of the hardware configuration. The MSIS has mainly been developed in C#, using Microsoft Visual Studio 2010 Ultimate as an integrated development environment (IDE). The dynamic link library (DLL) SPICEhelper.dll, which makes the required astrodynamical calculus using the NASA NAIF SPICE toolkit available in C#[4] has been developed in C++. The MSISF pattern preparation scripts are programmed in PHP 5. There are some additional supporting scripts written in MATLAB R2011b.

As said before, the MSISF makes use of or depends on other software components. If only the MSIS is to be used on the target machine, which is sufficient for the production of renderings, but not for the generation of surface patterns[5], the following components are mandatory:

- Microsoft .NET Framework 4.0

- POV-Ray for Windows 3.7 RC3[6]

---

[4] This thesis uses the NASA NAIF SPICE toolkit written in C/C++, which is also known as CSPICE. To make all CSPICE functions available in C#, a wrapper would be necessary, which goes beyond the thesis' scope. For that reason, all necessary astrodynamical calculations have been written in C++, compiled into a *.dll file and imported into the main C# project.

[5] This setup type is conceivable, if a pattern repository with all later used surface patterns is deployed along with the MSISF installation and there is no need to add/update surface patterns.

[6] A beta version (release candidate) of POV-Ray was used, because the 3.7 branch of POV-Ray will be the first one supporting multi-threaded rendering by default. Users with more than one CPU will profit from multi-threaded rendering.

Hardware Configuration Overview

Mainboard	ASUS P5Q Premium
CPU	Intel Core 2 Quad Processor Q6600 (8M Cache, 2.40 GHz, 1066 MHz FSB)
RAM	16 GB (4 × G.Skill DIMM 4 GB DDR2-800 F2-6400CL5D-8GBPQ)
GPUs	MSI nVIDIA GeForce GTX 260 896 MB GTX-260-T2D896-OC Palit nVidia GeForce GTS 450 Sonic 1024 MB DDR5
HDDs	OCZ Solid 3 2.5" SLD3-25SAT3-60G (System, RAID 1) Corsair CSSD-F60 60 GB (System, RAID 1) Samsung HD753LJ 750 GB (Data, RAID 1) WDC WD7500AADS-0 750 GB (Data, RAID 1)

Table 2.1 Hardware configuration for the machine used as test and development environment.

If all components of the MSISF should be used, the following additional software is necessary:

- MATLAB R2011b (The MathWorks) or MATLAB Compiler Runtime 7.16[7]

- MySQL 5.5.22

- PHP 5.3.10

To build the MSIS, a Microsoft Visual Studio 2010 version[8] and the full NASA NAIF SPICE Toolkit for C/C++ (CSPICE) N0064 is required; CSPICE has to be placed in the directory /src/SPICEhelper/CSPICE/ within the MSISF installation path.

For in-code documentation, doxygen 1.7.5 has been used as generator. The configuration file for a documentation rebuild after changes in the code documentation can be found in the source directory of the MSISF installation path as /src/Doxyfile.

---

[7] If MATLAB R2011b is not available on the target system, the MATLAB Compiler Runtime 7.16 can be used as runtime environment, without the need to install a full MATLAB system.
[8] The "Express" version of Microsoft Visual Studio 2010, which is available free of charge, might be sufficient for the compilation of the MSIS, but has not been tested.

2.3.2 MSIS User Interface, Definition of Inputs and Outputs

Before the software development starts, all possible modes of operation shall be defined for the main user interface, the MSIS. The MSIS is designed as a command-line application, which is controlled by passing arguments to the application through the command-line prompt. MSIS can operate in four distinct modes: The first one, for which the MSIS has primarily been designed, is the batch file processing mode. In this mode, the MSIS will open a user-specified text file, containing a set of so-called *fixed spacecraft states* per one text line. A spacecraft state defines the simulation time as modified Julian date (MJD) in UTC time, the spacecraft position within the ME/PA reference frame and the spacecraft orientation within this reference frame as rotation quaternion. This operation mode will generate one rendering per given state. By splitting large batch files, this operation mode can be used for the distributed computing of rendering sequences using more than one single machine.

The second operation mode is a single fixed state, which is given as a command-line argument instead of a batch file. This way, the MSIS will only generate one rendering per call. However, the MSIS can be invoked with multiple states using a standard Windows batch file (*.bat), which produces multiple renderings, but with a separate MSIS call for each rendering. This operation mode is mainly intended for testing purposes.

Both aforementioned operation modes require the user to pre-calculate a spacecraft position and orientation. If these parameters have not been pre-calculated, the user also has the possibility to call the MSIS with a specification of the spacecraft orbit around the Moon. Two separate possibilities have been implemented to describe an orbit: A set of Keplerian elements or a set of state vectors. In conjunction with a given epoch of these orbit parameters, the MSIS will automatically calculate the actual spacecraft position according to the actual simulation time. The MSIS internally works with Keplerian elements; given state vectors will be converted into a set of Keplerian elements.

With both of these operation modes, the MSIS can produce multiple renderings, if multiple simulation times are specified. Without specification of an orientation quaternion, the camera will be nadir-pointing; if an orientation quaternion is given, the camera will be rotated accordingly, but will orient to the same direction at every simulation time, unless an attitude transition (rotation rates about the camera's axes) is given.

For more details on the operation modes as well as a specification of the available command-line arguments, see section 6.3.

2.3.3 MSISF File System Layout

The MSISF can be installed to any user-given file system path. By default, the installer will place the MSISF installation in the standard path for program files, for example, `C:\Program Files\MSISF\`. The following list describes the directory structure of the MSISF installation:

`/`	MSISF installation directory (f.e. `C:\Program Files\MSISF\`)
`/COPYRIGHT.txt`	copyright information file
`/LICENSE.txt`	license information file
`/README.txt`	software information file
`/bin/`	binary files (compiled from the MSISF sources)
`/bin/batchconvert.exe`	generates a Windows *.bat file with single MSIS calls out of a MSIS batch file for distributed MSIS rendering
`/bin/delaunay2D.exe`	2D Delaunay Triangulation
`/bin/MSIS.exe`	Moon Surface Illumination Simulator (MSIS)
`/doc/`	documents
`/doc/MSIS_Code_Documentation/`	MSIS code documentation (HTML format, use `index.html`)
`/input/`	batchset input files for the MSIS
`/input/sample-scenarios.tab`	sample batch input file for the MSIS
`/install/`	MSISF and external installation files
`/kernels/`	SPICE kernels required by the MSIS
`/kernels/de421.bsp`	DE421 planetary and lunar ephemeris (binary SPK file)
`/kernels/moon_080317.tf`	lunar frame specifications for the DE421 ephemeris (text FK file)
`/kernels/moon_assoc_me.tf`	Moon ME/PA frame association kernel (text FK file)
`/kernels/moon_pa_de421_1900-2050.bpc`	high-accuracy lunar orientation data for the years 1900–2050 (binary PcK file)
`/kernels/naif0009.tls`	leapseconds kernel (text LSK file)
`/kernels/pck000009.tpc`	orientation, size and shape data for the natural bodies of the solar system (text PcK file)
`/lib/`	MSISF dynamic resources
`/lib/MSISRendering.dtd`	XML DTD for the rendering meta information XML file
`/lib/SPICEhelper.dll`	SPICE connector for the MSIS
`/output/`	MSIS output directory
`/pattern-repository/`	pattern repository
`/pattern-repository/4/`	surface patterns with 4 px/deg resolution
`/scripts/`	script files to be executed with other applications (PHP, MATLAB)
`/scripts/generate_pattern.php`	surface pattern generation script
`/scripts/ldem_mysql_insert.php`	NASA LRO LOLA LDEM MySQL import script
`/src/`	source code of the MSISF
`/src/Installer/`	NSIS installer source code for the MSISF
`/src/MSIS/`	source code for the MSIS
`/src/SPICEhelper/`	source code for `SPICEhelper.dll`
`/src/SPICEhelper/CSPICE/`	full installation of the NASA NAIF SPICE Toolkit for C/C++
`/src/delaunay2D.m`	MATLAB source file for `delaunay2D.exe`
`/src/Doxyfile`	configuration file for doxygen (in-code documentation)
`/src/MSISF.sln`	Microsoft Visual Studio 2010 solution file for the MSISF (containing the projects MSIS and SPICEhelper)

2.3.4 MSISF Deployment and Installation

The MSISF in its first version is non-publicly deployed as a compressed archive to the German Aerospace Center (DLR) along with compressed pattern repositories in resolutions of 4, 16 and 64 px/deg of the NASA LRO LDEM files in version 1.05.

Some time after graduation, the MSISF will be publicly deployed as a single setup file, which can be downloaded at http://go.rene-schwarz.com/masters-thesis; a setup wizard will guide the user through the installation process. This setup file contains a pattern repository in version 1.05 or greater of the NASA LOLA LDEM files with 4 px/deg resolution. This resolution should be sufficient for renderings at high flight altitudes and for trying the software. Pattern repositories with a greater resolution can be obtained separately from this website.

Those parts of the MSISF source code which have been written by the author will be subject to the conditions of the free and open source GNU General Public License (GNU GPL) version 2 or later. A complete release of the MSISF as a project on GitHub is envisaged.

Theoretical Foundations

3.1 The Mean Earth/Polar Axis Reference Frame

To coherently use spatial coordinates in the course of this thesis and the connected software development process, a definition of a reference frame to be used is required. As the Moon is the point of origin for all further work, it is natural to choose a body-fixed reference frame which is attached to the Moon. There are several definitions for such reference frames. However, the decision has been made to use the so-called *Mean Earth/Polar Axis reference frame* (abbreviated: ME/PA reference frame), which is consistent with the recommendations of the IAU/IAG Working Group on Cartographic Coordinates and Rotational Elements of the Planets and Satellites [6, p. 6]. In addition, the lunar digital elevation models (LDEMs) originating in the NASA Lunar Reconnaissance Orbiter (LRO) mission, which have been used for this thesis (see chapter 4), are also produced using this reference frame. All the following information has been obtained from [6].

The ME/PA reference frame defines a right-handed cartesian coordinate system, which is attached to the Moon's center of mass (COM) at its origin. The z-axis of this coordinate system is equivalent to the Moon's mean axis of rotation; the x-axis is normal to the z-axis and goes also through the Moon's COM. Additionally, the intersection point of the equator and the prime meridian is defined as being located on the x-axis. The y-axis is orthogonal to the plane defined by x- and z-axis and goes to the Moon's COM, too.

Last, the location of the prime meridian has to be declared to fix the coordinate system. In the ME/PA reference frame, the prime meridian is aligned in the so-called *mean Earth direction*. Because the Moon's movement is tidally locked to Earth, only one side of the Moon can be seen from Earth. By imagining a line between the Earth's COM and the Moon's COM, this

3 Theoretical Foundations

line will intersect the Moon's surface. This point of intersection is called *sub-Earth point*. This point moves during one Moon cycle due to the astrodynamical conditions[1]; however, the mean position of this point on the Moon's surface, called the *mean sub-Earth point*, is constant. The prime meridian is defined to cross this point.

Finally, the ME/PA reference frame also states planetocentric spherical coordinates, which will be called *selenographic[2] coordinates* in this thesis. As with the geographic coordinates on Earth, the single values of the selenographic coordinates are called *selenographic latitude* and *selenographic longitude*. The selenographic latitude is measured from the prime meridian, which marks 0° longitude, to the east (positive direction of rotation about the z-axis for the right-handed coordinate system) toward 360°. Alternatively, the longitude can be measured starting from the prime meridian to the east and to the west using the common +180° E/-180° W convention. The selenographic latitude is the angle between the equatorial plane and the vector from the COM to a point on the surface; it is measured from 0° for a point on the equator until +90° N/-90° S at both poles.

3.2 Derivation of a Spherical Coordinate System for the Moon

The Moon is commonly considered as a sphere; indeed it does not have the shape of a perfect sphere, but rather, has the shape of an oblate spheroid. However, the difference between its polar radius $r_{☾pol} = 1735.66$ km and mean equatorial radius $r_{☾eq} = 1738.64$ km is really small, i.e. its flattening is $f_☾ = 1 - (r_{☾pol}/r_{☾eq}) \approx 1/581.9$, so that a coordinate lattice can be modeled as a sphere, accepting a small misalignment in the z-axis for the conversion of selenographical coordinates into rectangular (cartesian) coordinates.

Given a sphere with an arbitrary radius r and center $\mathbf{O} = (0,0,0)^T$ in the origin of a spatial, right-handed, cartesian coordinate system with the coordinate triple $\mathbf{p} = (x,y,z)^T$ for an arbitrary point in the coordinate system, one can consider two distinct points $\mathbf{p}_1$ and $\mathbf{p}_2$ on the sphere's surface. It should be clear that it is difficult to declare rectangular coordinates with $\mathbf{p}_1 = (x_1, y_1, z_1)^T$, $\mathbf{p}_2 = (x_2, y_2, z_2)^T$ on the surface directly (cf. figure 3.1). Also the sphere eludes from an explicit, parametrized description so far.

By introducing an angle α between positive z-axis and the position vector $\mathbf{p}$ of an arbitrary point as well as an angle $\varphi \in \mathbb{R}$ between the positive x-axis and the position vector's projection onto the plane defined by x- and y-axis, the position of an arbitrary point on the sphere's

---

[1] A video of this behaviour prepared by NASA can be seen here: http://lunarscience.nasa.gov/articles/the-moons-hourly-appearance-in-2012/

[2] The term "selenographic" consists of the two Greek words Σελήνη (Moon) and γραφειν (to draw, to describe).

3.2 Derivation of a Spherical Coordinate System for the Moon

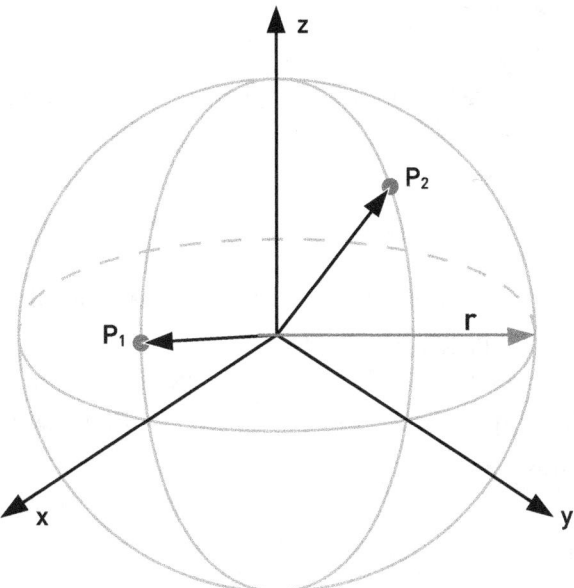

Figure 3.1 Two distinct points on the surface of a sphere with center in the coordinate origin $\mathbf{O} = (0, 0, 0)^{\mathrm{T}}$ and radius r. This illustration is based on the Wikipedia graphic http://de.wikipedia.org/wiki/Datei:Orthodrome_globe.svg, accessed on September 23, 2010.

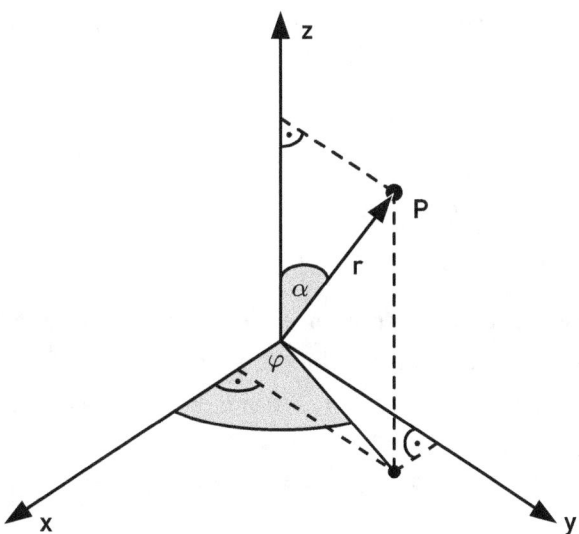

Figure 3.2 Representation of an arbitrary point **p** using spherical coordinates with polar angle α, azimuthal angle φ and the radial distance r to the orgin of the coordinate system.

surface can be described easily using the distance to the origin of the coordinate system and the two angles α and φ, where $\alpha \in [0, \pi]$ and $\varphi \in [0, 2\pi[$ by convention[3] (see figure 3.2).

In more general terms, a point at an arbitrary position in the coordinate system can now be represented using the polar angle α, azimuthal angle φ and the radial distance r to the orgin of the coordinate system:

$$p = (r, \alpha, \varphi) \tag{3.1}$$

This form of coordinates is referred to as spherical polar coordinates (or short: spherical coordinates). Considering only points on the sphere's surface, the distance to the origin of the coordinate system is the same for all points; it is the sphere's radius r. Now it is possible to describe an arbitrary point on the sphere's surface without being aware of its cartesian coordinates with two angles, since r is constant and known:

$$p = (\alpha, \varphi) \tag{3.2}$$

To convert spherical into rectangular coordinates and vice versa, a consideration of the relationship between spherical and rectangular coordinates is necessary. The radial distance r of one point is simply the EUCLIDean norm of its position vector $\mathbf{p}$:

$$r = \|\mathbf{r}\| = \sqrt{x^2 + y^2 + z^2} \tag{3.3}$$

By using trigonometric functions, it can be seen that the polar and azimuthal angle can be written as

$$\cos\alpha = \frac{z}{\sqrt{x^2 + y^2 + z^2}} \tag{3.4}$$

and

$$\tan\varphi = \frac{y}{x}. \tag{3.5}$$

With the help of these basic geometric considerations, the conversion from rectangular to spherical coordinates has been found. For the conversion from spherical to rectangular coordinates, some additional steps are necessary: From equation 3.4 follows

$$\cos\alpha = \frac{z}{\underbrace{\sqrt{x^2 + y^2 + z^2}}_{=r}}$$

$$\cos\alpha = \frac{z}{r} \longrightarrow \boxed{z = r\cos\alpha}. \tag{3.6}$$

---

[3] The angles α and φ are also called *polar angle* and *azimuth* in a spherical coordinate system.

The substitution of equation 3.3 by the equations 3.5 (reshaped to $x = \frac{y}{\tan \varphi}$) and 3.4 originates the relation

$$r = \sqrt{\frac{y^2}{\tan^2 \varphi} + y^2 + r^2 \cos^2 \alpha}$$

$$r^2 = y^2 \left(\frac{1}{\tan^2 \varphi} + 1 \right) + r^2 \cos^2 \alpha$$

$$y^2 = \frac{r^2 - r^2 \cos^2 \alpha}{\frac{1}{\tan^2 \varphi} + 1} \xrightarrow{\tan x = \frac{\sin x}{\cos x}} \frac{r^2 - r^2 \cos^2 \alpha}{\frac{\cos^2 \varphi}{\sin^2 \varphi} + 1}$$

$$= \frac{r^2 - r^2 \cos^2 \alpha}{\frac{\cos^2 \varphi + \sin^2 \varphi}{\sin^2 \varphi}} \xrightarrow{\cos^2 x + \sin^2 x = 1} (r^2 - r^2 \cos^2 \alpha) \sin^2 \varphi$$

$$= r^2 \sin^2 \varphi (1 - \cos^2 \alpha)$$

$$y = \sqrt{r^2 \sin^2 \varphi (1 - \cos^2 \alpha)}$$

$$= r \sin \varphi \cdot \sqrt{(1 - \cos^2 \alpha)} \xrightarrow{\pm\sqrt{1-\cos^2 x} = \sin x} r \sin \alpha \sin \varphi$$

$$\boxed{y = r \sin \alpha \sin \varphi}. \tag{3.7}$$

Using the reshape of equation 3.5 and the newly derived equation 3.7, x evaluates to

$$x = \frac{y}{\tan \varphi}$$

$$= \frac{r \sin \alpha \sin \varphi}{\tan \varphi} \xrightarrow{\tan x = \frac{\sin x}{\cos x}} \frac{r \sin \varphi \sin \alpha \cos \varphi}{\sin \varphi}$$

$$\boxed{x = r \sin \alpha \cos \varphi}. \tag{3.8}$$

The polar angle α used here is not identical to the selenographic latitude ϑ, since the selenographic latitude is measured starting with the equator, not the north pole. This way, $\vartheta = \alpha - \frac{\pi}{2}$. The conversion from selenographic coordinates (latitude ϑ and longitude φ) to rectangular coordinates within the ME/PA reference frame can now be written as

$$\mathbf{p}(\vartheta, \varphi) = \begin{pmatrix} p_x(\vartheta, \varphi) \\ p_y(\vartheta, \varphi) \\ p_z(\vartheta, \varphi) \end{pmatrix} = \begin{pmatrix} r \cos \vartheta \cos \varphi \\ r \cos \vartheta \sin \varphi \\ r \sin \vartheta \end{pmatrix} \tag{3.9}$$

and vice versa as

$$p(x, y, z) = p(\vartheta(x, y, z), \varphi(x, y, z)) = \left(\arcsin \frac{z}{r}, \arctan2(y, x) \right). \tag{3.10}$$

4 CHAPTER

Creating a Global Lunar Topographic Database

4.1 Overview of Available Lunar Topographic Data

To build a spatial digital elevation model of the Moon's surface, an investigation of available topographic data for the Moon is necessary. Four publicly available data sets of the Moon's topography have been identified for the purpose of this thesis during the investigation.

The first and oldest available data set originated in the NASA **Clementine mission** (officially called *Deep Space Program Science Experiment* (DSPSE)) in 1994. The topographic data is retrievable as a global equally gridded map at the finest resolution[1] of 4 px/deg. The given height differences are relative[2] to a spheroid with a radius of 1 738 km at the equator and a flattening of 1/3234.93. Thus, a resolution of $\approx$ 7.58 km/px at the equator is achieved.

A tentative attempt revealed that this data is not precise enough for the thesis' scope. Figure 4.1 shows a visualization of the Clementine topographic data on a plane, where it can be seen that the spatial resolution is too low for a realistic surface simulation. This is aggravated by the

---

Chapter Image: NASA's Lunar Reconnaissance Orbiter (LRO) and NASA's Lunar Crater Observation and Sensing Satellite (LCROSS), ready for liftoff on an Atlas V/Centaur rocket at Cape Canaveral Air Force Station in Florida on June 18, 2009. ©2009 NASA/Ken Thornsley, License: Public Domain. Available at http://mediaarchive.ksc.nasa.gov/detail.cfm?mediaid=41893.

[1] The Clementine gravity and topography data products are described here: http://pds-geosciences.wustl.edu/lunar/clem1-gravity-topo-v1/cl_8xxx/aareadme.txt

[2] See the PDS label file for the finest resolution: http://pds-geosciences.wustl.edu/lunar/clem1-gravity-topo-v1/cl_8xxx/topo/topogrd2.lbl

4 Creating a Global Lunar Topographic Database

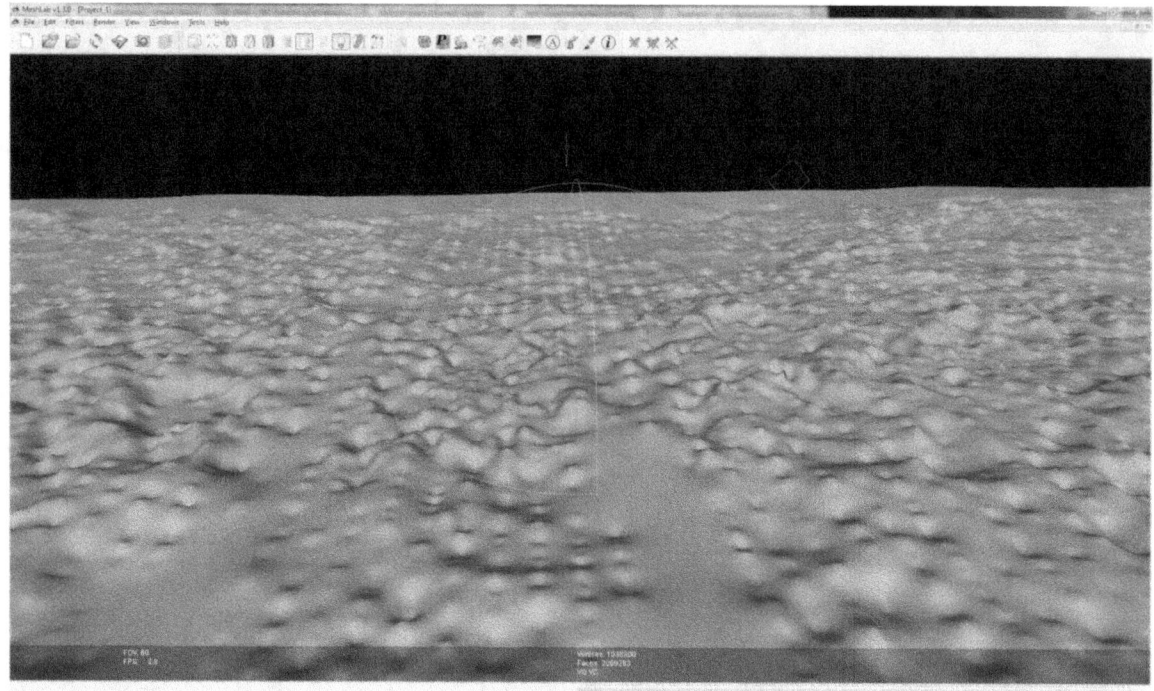

Figure 4.1 Plane visualization of the Clementine topographic data at the finest resolution (4 px/deg). A visual examination of the data shows that this data set is impractical to generate a realistic surface illumination.

fact that some data grid points seem to be interpolated, resulting in a non-natural transition between two sample points.

Two more topographic data sets are available from the 2007 JAXA SELENE (*Selenological and Engineering Explorer*) mission and its **Kaguya** lunar orbiter spacecraft. First, a DTM is available from the LALT (*Laser Altimeter*) with a resolution of 16 px/deg. This resolution is sufficient for a global or high-altitude rendering of the Moon's surface, but does not meet the requirements for a surface close-up rendering. In addition, DEMs generated out of stereo images from the *Terrain Camera* (TC) are available with a resolution of 4 096 px/deg, which can be retrieved using the SELENE Data Archive[3]. This data archive offers many options for the post-processing of the SELENE data products, for example, the alteration of the map projection (equirectangular, MERCATOR's, orthographic, polar stereographic, LAMBERT's conformal conic or sinusoidal projection) as well as a resolution reduction from 4 096 px/deg to 2 048, 1 024, 512, 256, 128, 64 and resolutions <64 px/deg using distinct interpolation methods (nearest neighbor,

---

[3] The SELENE Data Archive can be accessed at https://www.soac.selene.isas.jaxa.jp/archive/index.html.en.

bi-linear, cubic convolution). Unfortunately, the DEM is split into 7 200 separate files with a file size of about 2.2 TiB; the data archive only allows the download of 100 files with a maximum file size of 3 GiB simultaneously. For the thesis' purpose, a resolution of 128 px/deg should be sufficient in most cases. To retrieve a global DEM with such a resolution, laborious steps regarding the post-processing, retrieval and data conversion need to be carried out, which are beyond the thesis' time limits.

A fourth source of a lunar DEM is the NASA **Lunar Reconnaissance Orbiter** (LRO) spacecraft, which was launched in 2009. The *Lunar Orbiter Laser Altimeter* (LOLA) instrument aboard the LRO can produce DEMs with a maximum resolution of 1 024 px/deg ($\approx$ 29.612 m/px) [107, p. 239, table 9]. DEMs (so-called "products" or "LRO Lunar Digital Elevation Models" — LRO LDEMs) will be offered in eight resolutions: 4, 16, 32, 64, 128, 256, 512 and 1 024 px/deg [107, op. cit.]. These DEMs, their resolutions and their file formats are ideally suited for the thesis' purposes.

4.2 Data from NASA's Lunar Orbiter Laser Altimeter (LOLA)

The Lunar Orbiter Laser Altimeter (LOLA) is an instrument aboard NASA's Lunar Reconnaissance Orbiter (LRO), which is designed to gain data about Moon's surface, supporting the selection of landing sites for future exploration missions [107, p. 210]. The primary objective of LOLA is the development of an accurate (approx. 50 m) global geodetic grid [107, p. 212]. Other goals of the LOLA instrument are the characterization of the illumination conditions of the polar region, the imaging of permanently shadowed regions, the identification of surface polar ice, if present, [83, p. 322] ascertaining what goes along with the determination of the surface reflectance[4], as well as geodetic location, direction and magnitude of surface slopes [107, p. 210].

LOLA uses laser altimetry as a measuring principle: First, an accurate range measure from the LRO to the lunar surface is done. In conjunction with a precise LRO orbit determination, a referencing of the surface ranges to the Moon's center of mass is possible. This way, a high-precision global geodetic grid in the ME/PA reference frame can be gathered. For this purpose, LOLA has five laser beams (cf. picture 4.3) operating at a fixed rate of 28 Hz and an overall accuracy of 10 cm. [107, p. 209]. This means there is one shot for approximately every 57 m for a nominal ground track velocity of 1 600 m/s, while each laser footprint ("spot") will have a diameter of 5 m at a nominal flight altitude of 50 km [102, p. 4]. The spots are arranged

---

[4] The presence of a critical amount of water ice crystals on the Moon's surface would result in a measurable increase in the surface reflectance [107, p. 210].

4 Creating a Global Lunar Topographic Database

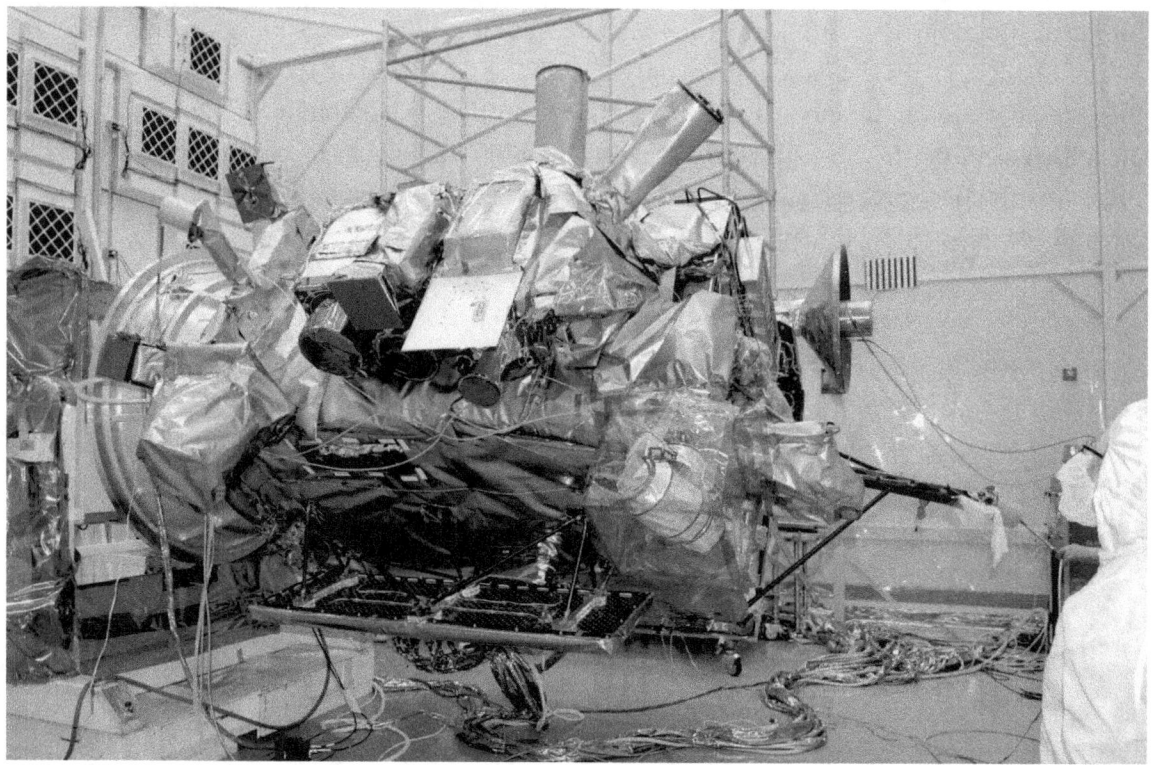

Figure 4.2 The Lunar Reconnaissance Orbiter (LRO) in a near-final construction stage. The entire instrument suite is visible from this perspective; the Lunar Orbiter Laser Altimeter (LOLA) is the conical instrument directly beyond the white shining plate. © NASA/Debbie McCallum. Obtained from http://www.nasa.gov/mission_pages/LRO/multimedia/lrocraft5.html.

in the form of a cross, being 25 m apart from each other; the cross is being rotated by 26° counterclockwise to the prograde movement of the spacecraft, preferably to achieve a high level of coverage [102, p. 4].

Several data products are generated using the LOLA data:

- EDR (Experiment Data Records): Raw, uncalibrated data from the LOLA instrument.

- RDR (Reduced Data Records): Calibrated, geolocated pulse returns, altitudes and reflectivities. This information is produced after a range calibration and orbital processing using information of the spacecraft's trajectory, attitude history and a lunar orientation model. All higher-level products are generated from the cumulative RDR product.

- SHADR (Spherical Harmonic Analysis Data Records): SHADRs contain coefficients for a lunar spherical harmonic model of the lunar shape. They will be generated out of GDRs.

4.2 Data from NASA's Lunar Orbiter Laser Altimeter (LOLA)

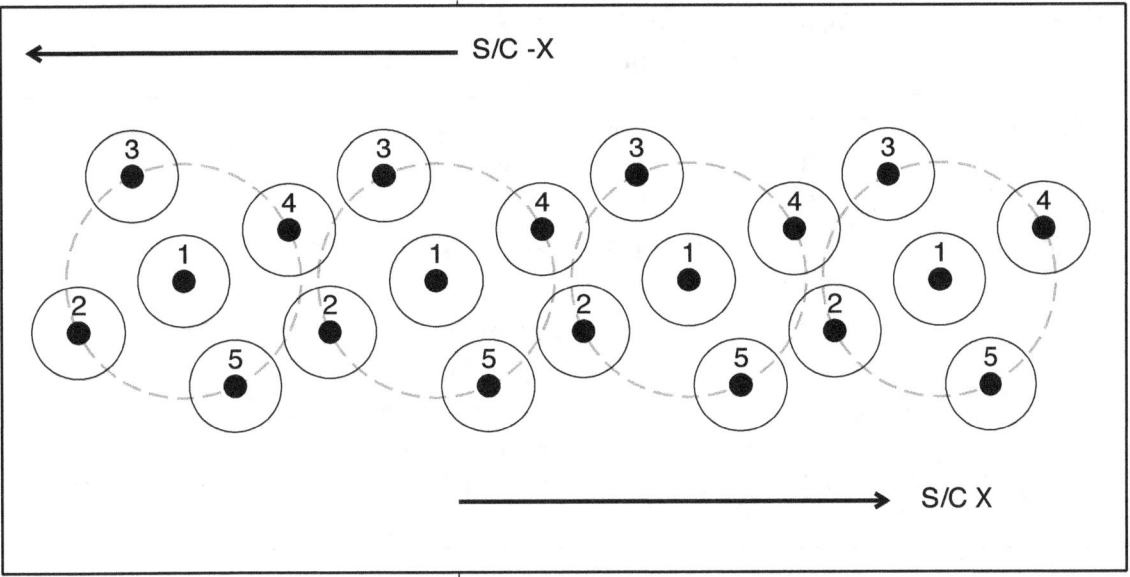

Figure 4.3 LOLA laser geometry on the ground for four consecutive shots; the numbers represent the channel numbers. The solid, black-filled circles indicate the transmitted laser footprints on the Moon's surface, while the solid, concentric circles imply the receiver's field of view. © NASA. Obtained from http://lunar.gsfc.nasa.gov/lola/images/fig.pdf.

- GDR (Gridded Data Records): GDRs are the primary products of the LOLA instrument. They are raster digital elevation models (DEMs) of the lunar radius with reference to a spherical reference point about the Moon's center of mass. GDRs are available in different resolutions (see table 4.1).

The GDRs are also called *Lunar Digital Elevation Models* (LDEMs), which differ not only in terms of available resolution, but also projection method used. Global LDEMs are offered in an equi-rectangular map projection (simple cylindrical projection), while the polar products, which are usually in a higher resolution, are made available in polar stereographic projection. [107, pp. 236 ff.].

LDEMs are formatted as binary pictures of the PDS[5] object type IMAGE. They are also provided in a geo-referenced JPG-2000 format. [107, p. 239].

A PDS object of the IMAGE type is a two-dimensional array of values, stored in a binary file. The values are all of the same data type and each value is referred to as *sample*. An IMAGE object

---

[5] The PDS format is a standardized data format for the distrubution of scientific mission data across all NASA missions. PDS stands for *Planetary Data System*, which is the NASA data distribution system for NASA's planetary missions.

Product	Size	Resolution in $\frac{m}{px}$	Tiles
LDEM_4	4 MiB	7 580.8	— (global)
LDEM_16	64 MiB	1 895	— (global)
LDEM_32	256 MiB	947.6	— (global)
LDEM_64	1 GiB	473.8	— (global)
LDEM_128	2 GiB	236.9	— (global)
LDEM_256	4×2 GiB	118.45	4 tiles, longitudes 0:180:360 by northern/southern hemisphere
LDEM_512	16×2 GiB	59.225	16 tiles, longitudes in 45° segments by northern/southern hemisphere
LDEM_1024	64×2 GiB	29.612	64 tiles, longitudes in 45° segments, latitude in 22.5° segements

Table 4.1 Available LRO LOLA equi-rectangular map-projected LDEM products. Based on [107, p. 239].

consists of a series of lines, which contain a fixed number of samples per line. Each binary file has a detached meta information file, called PDS label (*.lbl). The PDS Standard [76] requires the four essential parameters

- LINES: number of lines in the image
- LINE_SAMPLES: number of samples in each line
- SAMPLE_BITS: number of bits in each individual sample
- SAMPLE_TYPE: data type of the samples

to be defined for an IMAGE-type object in a PDS label file. [76, p. A-64].

Usually, a LDEM PDS label file contains a lot more information and parameters, each of which is discussed in the course of this chapter. This thesis makes use of version 1.05 of the LRO LOLA data, which was released on March 15, 2011. According to the NASA statements in the errata file[6], these data products are only preliminary; they will be updated regularly with new and more precise data, as the mission progresses. In other words, the used LDEM products are partially interpolated, which introduces visible errors in the later MSIS renderings (see figure 4.4 for an example rendering with visible flaws in the surface data caused by data interpolation).

---

[6] available at http://imbrium.mit.edu/ERRATA.TXT

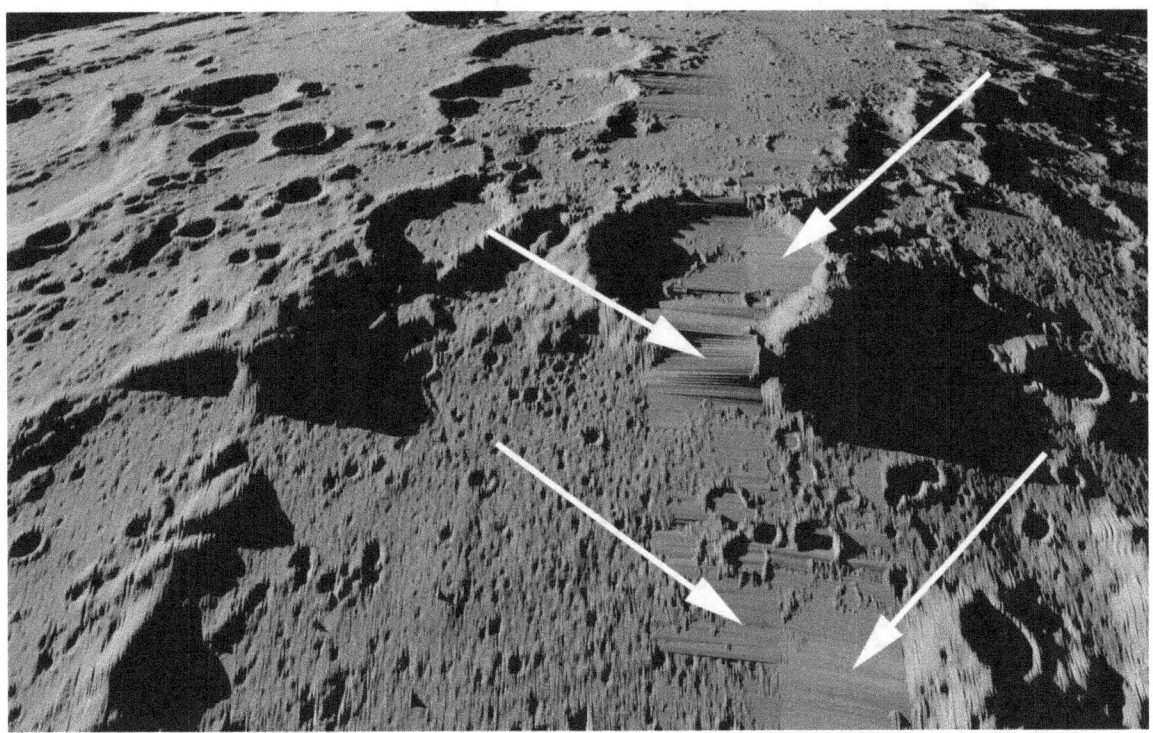

Figure 4.4 Visible flaws in the surface data caused by data interpolation in the preliminary LRO LOLA LDEM products, shown on an example MSIS rendering.

For version 1.05 of the LOLA LDEM data, NASA states that the LDEM_64 product, which is the highest resolution to be used in this thesis, is 45 % sampled and 55 % interpolated. At the time of completion of this thesis, NASA released version 1.09 of the LOLA data on March 15, 2012, but the last statement concerning the data accuracy was made for version 1.07, released on September 15, 2011. Table 4.2 gives a comparison of the data quality of versions 1.05 and 1.07.

4.3 LOLA Data Import and Conditioning

4.3.1 Overall Concept

Since the LOLA LDEM products are all of the same scheme, the usage of a database system has been identified as the most efficient way to handle and store data. For this purpose, MySQL, a free and open source database system, will be used. A MySQL database will maintain all information regarding the lunar topography with additional data. In order to import the data from the LOLA LDEM products, data processing is necessary.

4 Creating a Global Lunar Topographic Database

LDEM product	Version 1.05		Version 1.07	
	fraction sampled	fraction interpolated	fraction sampled	fraction interpolated
LDEM_4	$\approx 100\%$	$\approx 0\%$ (only 33 bins lack shots)	100 %	0 %
LDEM_16	90 %	10 %	$\approx 100\%$	$\approx 0\%$
LDEM_64	45 %	55 %	55 %	45 %
LDEM_128	28 %	72 %	35 %	65 %
LDEM_256	17 %	83 %	21 %	79 %
LDEM_512	–	–	13 %	87 %

Table 4.2 Data coverage of the LOLA LDEM products for versions 1.05 and 1.07. Source: Errata file of the LOLA PDS data node at http://imbrium.mit.edu/ERRATA.TXT.

This data processing and conditioning could be done with many programming languages and concepts. To achieve rapid development, the scripting language PHP has been used, because data processing is possible using only a few lines of code, the code is simple to understand and PHP contains native functions for the connection and querying of MySQL databases.

First, considerations regarding the database and table design, as well as the MySQL server configuration are necessary; these considerations are discussed in the next section. Subsequently, the process of the data import and conditioning using the MSISF LDEM MySQL import script `ldem_mysql_insert.php` is elucidated in detail.

4.3.2 MySQL Database Design, Query Optimization and Commitment of the MySQL Server Configuration to Database Performance

To store millions of data records[7] regarding a global lunar topography, preparatory considerations about the database design have to be made, since a well-conceived database design will have a serious impact on the performance of database queries — a fact that shall be illustrated in this section.

MySQL supports multiple databases per server instance, while each database can consist of one or more tables together with corresponding views, triggers and procedures. A database can be thought of as containing a myriad of datasets to a particular topic. For the purpose of the MSISF, just one database, storing the lunar topography, is required. A separate table for

---

[7] The LDEM_64 dataset, the biggest resolution to be used in the MSISF, will have a spatial resolution of 64 px/deg latitude and longitude, which means that $64^2 \cdot 360 \cdot 180 = 265\,420\,800$ samples have to be stored.

each available LRO LDEM resolution has been found to be appropriate with respect to arbitrary expandability, data separation, storage requirements and query optimization[8].

Furthermore, MySQL offers several possibilities for database-internal table structuring and storage, which results in the availability of several storage engines (e.g. MyISAM, InnoDB, CSV, Archive, etc. pp.). As of MySQL 5.5.5, InnoDB is the default storage engine for MySQL tables, because InnoDB is ACID[9]-compliant and has advanced features concerning commit, rollback, crash-recovery and row-level locking capabilities. MyISAM is a little bit older than InnoDB, lacking the aforementioned features, but is a well-proven storage engine, which is mainly used for web applications and data warehousing. [94]

Nevertheless, MyISAM remains the only storage engine capable of spatial indices in spatial columns [95]. This feature allows a severe acceleration of spatial queries to the database, as illustrated later. After these considerations, one database with single tables using the MyISAM storage engine for each LRO LDEM resolution will be used. Now, the table-level design remains to be dealt with. A reasonable starting point for these deliberations is an examination of the raw LRO LOLA LDEM data.

The LOLA LDEM data sets deliver one piece of information explicitly, the height over MMR for a certain pair of latitude and longitude, which themselves can be inferred by the byte position of the height information in the LDEM file. So latitude and longitude are both defined implicitly within the LRO LDEM files. An ingenuous attempt to create a first table structure could be this MySQL table creation statement:

```
CREATE TABLE IF NOT EXISTS 'ldem'
(
  'lat' double NOT NULL,
  'lon' double NOT NULL,
  'height' double NOT NULL,
  PRIMARY KEY ('lat','lon')
) ENGINE=MyISAM DEFAULT CHARSET=utf8;
```

This table design stores all DEM data in a simple key/value form, whereas the primary key is a multi-column index of the latitude and longitude value, and does not utilize MySQL's spatial extensions. To retrieve all elevation information within a rectangular surface patch, starting at 25° N 10° E in the upper left corner and ending at 30° N 15° E in the bottom right corner, one can use the following MySQL query:

---

[8] Another approach could be the storage of all resolutions in one single table, referencing the respective resolution of one record against a foreign key of a resolution table. This method is called database normalization, but should perform poorly for the designated purpose.

[9] ACID is an acronym for <u>A</u>tomicity, <u>C</u>onsistency, <u>I</u>solation and <u>D</u>urability, a concept for the reliable processing of database transactions.

4 Creating a Global Lunar Topographic Database

```sql
SELECT height FROM ldem
WHERE
    lat >= 25 AND lon >= 10
    AND lat < 30 AND lon < 15
ORDER BY lat, lon;
```

MySQL performs slowly on this query, especially with increasing resolution. Table 4.3 shows a MySQL query performance comparison for this query for the three LDEM resolutions used.

A profiling[10] of the MySQL query reveals that nearly all query time (450.910 s out of 451.554 s total) is spent in executing the query (cf. table 4.4), while all other processing times for subsequent processes such as sorting, caching, and logging are negligible.

Furthermore, the table design demands the later user to have knowledge of the used reference radius of the Moon, to which all height values are relative. In the case of the LRO LDEMs this is an arbitrarily chosen reference radius of 1 737.4 km, while the current best estimate of the volumetric mean Moon radius is $1.73715 \cdot 10^6$ (± 10) m [81, p. 898]. An inexperienced user could introduce a systematic error of 250 (± 10) m to all radius queries. The best practice should be to provide an unambiguous and intuitive table design, which should provide additional pre-processed data, for example, the rectangular coordinates of the respective surface point in the ME/PA reference frame, making subsequent calculations superfluous.

As said before, a regular query will perform poorly on the database, because no specialized data index has been set up. MySQL's MyISAM storage engine is capable of spatial indices in spatial columns; it is obvious to create a spatial index out of a latitude/longitude pair. A spatial location is stored as value of the data type point in MySQL, which is simply constructed using the values of the latitude and longitude. Subsequently, the resulting column is marked as SPATIAL KEY, which is a data index for the table. The following MySQL code snippet shows the intermediate table layout for the development, which was used for each LDEM resolution (shown for LDEM_4 as example):

```sql
CREATE TABLE IF NOT EXISTS 'ldem_4' (
  'point_id' bigint(20) unsigned NOT NULL AUTO_INCREMENT COMMENT 'point ID',
  'line' int(10) unsigned NOT NULL,
  'sample' int(10) unsigned NOT NULL,
  'lat' double NOT NULL COMMENT 'latitude [deg]',
  'lon' double NOT NULL COMMENT 'longitude [deg]',
  'height' double NOT NULL,
```

---

[10] The profiling feature is a community feature of MySQL. The output is shifted line-by-line due to a programming error. This means, for example, the time 443.645335 s (column "Duration") in table 4.4 belongs to the process "executing", not to "sorting result". This is a known bug of MySQL (bug #52492: "Issue with output of SHOW PROFILE", http://bugs.mysql.com/bug.php?id=52492).

	LDEM_4	LDEM_16	LDEM_64
number of affected rows	400	6 400	102 400
query time	1.723 s	27.421 s	451.554 s

Table 4.3 MySQL query performance comparison for a 5° × 5° surface patch using standard indices (`PRIMARY KEY(...)`).

Status	Duration	CPU_user	CPU_system
starting	0.000023	0.000000	0.000000
waiting for query cache lock	0.000013	0.000000	0.000000
checking query cache for query	0.000157	0.000000	0.000000
checking permissions	0.000013	0.000000	0.000000
opening tables	0.202970	0.000000	0.000000
system lock	0.000013	0.000000	0.000000
waiting for query cache lock	0.000030	0.000000	0.000000
init	0.000116	0.000000	0.000000
optimizing	0.000029	0.000000	0.000000
statistics	0.000026	0.000000	0.000000
preparing	0.000052	0.000000	0.000000
executing	0.000004	0.000000	0.000000
sorting result	450.910301	80.434116	48.375910
sending data	0.027119	0.000000	0.000000
waiting for query cache lock	0.000013	0.000000	0.000000
sending data	0.001940	0.000000	0.000000
waiting for query cache lock	0.000014	0.000000	0.000000
...	...	...	...
waiting for query cache lock	0.000014	0.000000	0.000000
sending data	0.002939	0.000000	0.000000
waiting for query cache lock	0.000021	0.000000	0.000000
sending data	0.047889	0.000000	0.000000
end	0.000012	0.000000	0.000000
query end	0.000004	0.000000	0.000000
closing tables	0.000017	0.000000	0.000000
freeing items	0.002088	0.000000	0.000000
logging slow query	0.000008	0.000000	0.000000
logging slow query	0.000008	0.000000	0.000000
cleaning up	0.000007	0.000000	0.000000

Table 4.4 MySQL profiling for the slow query on the LDEM_64 table (all values in seconds).

4 Creating a Global Lunar Topographic Database

```
 8    'planetary_radius' double NOT NULL,
 9    'x' double DEFAULT NULL,
10    'y' double DEFAULT NULL,
11    'z' double DEFAULT NULL,
12    'point' point NOT NULL,
13    PRIMARY KEY ('point_id'),
14    SPATIAL KEY 'point_index' ('point')
15  ) ENGINE=MyISAM DEFAULT CHARSET=utf8 AUTO_INCREMENT=1;
```

The table columns `line`, `sample` and `height` have been used for verification purposes only during development; in the final table layout they have been removed to reduce the table storage requirements. The final table layout is given by the following MySQL statement (again on the example of LDEM_4):

```
 1  CREATE TABLE IF NOT EXISTS 'ldem_4' (
 2    'point_id' bigint(20) unsigned NOT NULL AUTO_INCREMENT COMMENT 'point ID',
 3    'lat' double NOT NULL COMMENT 'latitude [deg]',
 4    'lon' double NOT NULL COMMENT 'longitude [deg]',
 5    'planetary_radius' double NOT NULL,
 6    'x' double DEFAULT NULL,
 7    'y' double DEFAULT NULL,
 8    'z' double DEFAULT NULL,
 9    'point' point NOT NULL,
10    PRIMARY KEY ('point_id'),
11    SPATIAL KEY 'point_index' ('point')
12  ) ENGINE=MyISAM DEFAULT CHARSET=utf8 AUTO_INCREMENT=1;
```

Listing 4.1 Final table layout.

A comparison of the storage requirements between the intermediate table layout and the final table layout shows (see table 4.5), that additional columns of simple data types, like `int` and `double`, do not have a big influence on the table size. Noting this, the column `planetary_radius` was kept for conceivable further uses of the spatial database.

An optimized spatial query for the example $5° \times 5°$ surface patch, starting at $25°\,N\ 10°\,E$ in the upper left corner and ending at $30°\,N\ 15°\,E$ in the bottom right corner, utilizing MySQL spatial extensions, can now be written as:

```
1  SELECT planetary_radius FROM ldem
2  WHERE
3      MBRContains(GeomFromText('POLYGON((25 10, 25 15, 30 15, 30 10, 25 10))'), point)
4  ORDER BY lat, lon;
```

4.3 LOLA Data Import and Conditioning

	LDEM_4		LDEM_16		LDEM_64	
	interm.	final	interm.	final	interm.	final
number of data rows	1 036 800		16 588 800		265 420 800	
average row size	248 B	232 B	250 B	234 B	250 B	234 B
space usage for data	106.8 MiB	91.0 MiB	1 708.6 MiB	1 455.5 MiB	27 337.1 MiB	23 287.5 MiB
space usage for index	138.1 MiB	138.1 MiB	2 240.4 MiB	2 240.4 MiB	35 989.3 MiB	35 989.3 MiB
total table size	244.9 MiB	229.0 MiB	3 949.0 MiB	3 695.9 MiB	63 326.5 MiB	59 276.8 MiB

Table 4.5 Comparison of the storage requirements for the used LDEM resolutions distinguished by the intermediate and the final table layout.

This query makes use of a so-called *minimum bounding rectangle* (MBR), the smallest possible axes-parallel rectangle, enclosing a defined set of points or objects in a two-dimensional space. Within the thesis' scope it is the minimum $(\min(\vartheta), \min(\varphi))$ and maximum $(\max(\vartheta), \max(\varphi))$ extent of the latitude and longtitude of a defined set of points. `MBRContains` is one kind of MySQL's spatial functions for testing spatial relations between geometric objects. In combination with the spatial index, which results internally in an R-tree[11] using the MyISAM storage engine [92], highly optimized execution algorithms can be used. Spatial indices allow the effective computation of questions regarding measurements and relations between geometrical objects, in contrast to the traditional indices of non-spatial databases, which are unable to handle such queries effectively.

A performance comparison (see table 4.6) shows that up to 93.5 % of the overall query execution time can be saved using a spatial data index. The query profiling (cf. table 4.7) reveals that the time savings are achieved during the execution process, as intended.

Here, it is important to mention that the MySQL server configuration has a significant impact on the execution time of the SQL query. During tests for a reasonable configuration, execution times up to 45 minutes (not using MySQL spatial extensions) occurred. The aforementioned execution times could be achieved using an optimized MySQL server instance configuration. The MySQL server instance configuration file (`*.ini` file) is attached as appendix B.1.

---

[11] In computer science, R-trees are multi-dimensional index structures, which are organized in a tree data structure. An R-tree requires more memory and causes higher computation times for changes in the indexed data, but it has a high spatial query performance. In the case of this thesis, the database will change rarely, so the query time is the most important property.

	LDEM_4	LDEM_16	LDEM_64
number of affected rows	400	6 400	102 400
query time	0.297 s	2.133 s	29.342 s
time savings compared to standard method	1.426 s (-82.8 %)	25.288 s (-92.2 %)	422.212 s (-93.5 %)

Table 4.6 MySQL query performance comparison for a 5° × 5° surface patch using a spatial index (SPATIAL KEY ...).

Status	Duration	CPU_user	CPU_system
starting	0.000019	0.000000	0.000000
waiting for query cache lock	0.000012	0.000000	0.000000
checking query cache for query	0.000173	0.000000	0.000000
checking permissions	0.000011	0.000000	0.000000
opening tables	0.012992	0.000000	0.000000
system lock	0.000031	0.000000	0.000000
init	0.000068	0.000000	0.000000
optimizing	0.000018	0.000000	0.000000
statistics	0.029689	0.000000	0.000000
preparing	0.000055	0.000000	0.000000
executing	0.000004	0.000000	0.000000
sorting result	28.276892	0.405603	1.560010
sending data	0.957158	0.234002	0.702005
end	0.000011	0.000000	0.000000
query end	0.000002	0.000000	0.000000
closing tables	0.000013	0.000000	0.000000
freeing items	0.000336	0.000000	0.000000
logging slow query	0.000004	0.000000	0.000000
logging slow query	0.000003	0.000000	0.000000
cleaning up	0.000003	0.000000	0.000000

Table 4.7 MySQL profiling for the optimized query on the LDEM_64 table (all values in seconds).

4.3.3 Importing the LOLA Data into the MySQL Database

The import process of a NASA LRO LDEM file is done via the ldem_mysql_insert.php script (located in the MSISF-directory /scripts; see appendix B.3 for a printed version), which can be run from the windows command prompt[12] as PHP script:

```
php ldem_mysql_insert.php
```

The configuration variables at the beginning of the script have to be edited before script execution. There are six necessary variables, which need to be adjusted to the actual use case:

```
76  # ==========================================================
77  # Configuration of the MySQL import process
78  # ==========================================================
79
80  # LDEM resolution (integer) to be used (preconfigured for LDEM_4, LDEM_16,
81  # LDEM_64, LDEM_128, LDEM_256, LDEM_512 and LDEM_1024).
82  $LDEM = 1024;
83
84  # Path to LDEM directory, where all LDEM files (*.img) to be imported are placed
85  # (w/o trailing slash)
86  $path_LDEM = "L:/path/to/LDEM";
87
88  # MySQL connection parameters
89  $mysql_host = "localhost";      # MySQL server, e.g. "localhost" or "localhost:3306"
90  $mysql_db   = "";               # MySQL database
91  $mysql_user = "";               # MySQL username for specified database
92  $mysql_pwd  = "";               # MySQL password for specified username
```

$LDEM defines the LRO LDEM resolution that is going to be imported; $path_LDEM denotes the file system path of the LDEM file. It is important to ensure that the original file names from NASA are kept, since the script will make use of the NASA file name scheme. The other four variables configure the connection to the MySQL database with their intuitive meanings ($mysql_host, $mysql_db, $mysql_user, $mysql_pwd). The import script expects the existence of the corresponding LDEM_... table[13] with the final table layout as specified as in the previous section; this table has to be created manually before script execution.

---

[12] It is assumed that a recent version of PHP (http://php.net) is properly installed and the PHP binary php.exe is available through the system's PATH variable.

[13] The table for a LDEM with a resolution of, for example, 64 px/deg, has to be named LDEM_64.

Additionally, the import script needs some information about the LDEM file itself to ensure the correct allocation of the spatial position for each single byte[14]. The corresponding configuration variables have been prepared for each available LDEM resolution, that is 4, 16, 64, 128, 256, 512 and 1 024 px/deg. If new LDEM versions are released by NASA in future, these values could change, so a manual verification before using new LDEM files is advised.

Each LDEM file has a corresponding label file (*.lbl), in which all information regarding the binary LDEM file is stored in the PDS format. A PDS file is a simple text file with a standardized syntax. The required parameters are

- the LDEM resolution (MAP_RESOLUTION),
- the number of data samples per line (LINE_SAMPLES),
- the line containing the last pixel (LINE_LAST_PIXEL),
- the line projection offset (LINE_PROJECTION_OFFSET) and
- the sample projection offset (SAMPLE_PROJECTION_OFFSET).

The values of the specified parameters have been inherited from the current available LDEM files with version 1.05 (2011/03/15) and are stored in the corresponding configuration variables:

```
# LDEM-specific options
# These parameters can be found in the corresponding *.lbl files for each *.img file.
# Preconfigured for LDEM_4, LDEM_16, LDEM_64, LDEM_128, LDEM_256, LDEM_512 and LDEM_1024.
switch($LDEM)
{
    case 4:
    default:
    /* LDEM_4 */
        $c_MAP_RESOLUTION = 4;
        $c_LINE_SAMPLES = 1440;
        $c_LINE_LAST_PIXEL = 720;
        $c_LINE_PROJECTION_OFFSET = 359.5;
        $c_SAMPLE_PROJECTION_OFFSET = 719.5;
        break;

    case 16:
    /* LDEM_16 */
        $c_MAP_RESOLUTION = 16;
        $c_LINE_SAMPLES = 5760;
```

---

[14] The import script is designed for the import of LDEM files in simple cylindrical projection only. These LDEM data files can be retrieved from http://imbrium.mit.edu/DATA/LOLA_GDR/CYLINDRICAL/IMG/. Other files differ in the projection type used and therefore need to be processed in a different way.

```
            $c_LINE_LAST_PIXEL = 2880;
            $c_LINE_PROJECTION_OFFSET = 1439.5;
            $c_SAMPLE_PROJECTION_OFFSET = 2879.5;
            break;

        case 64:
        /* LDEM_64 */
            $c_MAP_RESOLUTION = 64;
            $c_LINE_SAMPLES = 23040;
            $c_LINE_LAST_PIXEL = 11520;
            $c_LINE_PROJECTION_OFFSET = 5759.5;
            $c_SAMPLE_PROJECTION_OFFSET = 11519.5;
            break;

        case 128:
        /* LDEM_128 */
            $c_MAP_RESOLUTION = 128;
            $c_LINE_SAMPLES = 46080;
            $c_LINE_LAST_PIXEL = 23040;
            $c_LINE_PROJECTION_OFFSET = 11519.5;
            $c_SAMPLE_PROJECTION_OFFSET = 23039.5;
            break;

        case 256:
        /* LDEM_256 */
            $c_MAP_RESOLUTION = 256;
            $c_LINE_SAMPLES = 46080;
            $c_LINE_LAST_PIXEL = 23040;
            break;

        case 512:
        /* LDEM_512 */
            $c_MAP_RESOLUTION = 512;
            $c_LINE_SAMPLES = 46080;
            $c_LINE_LAST_PIXEL = 23040;
            break;

        case 1024:
        /* LDEM_1024 */
            $c_MAP_RESOLUTION = 1024;
            $c_LINE_SAMPLES = 30720;
            $c_LINE_LAST_PIXEL = 15360;
            break;
}
```

4 Creating a Global Lunar Topographic Database

For LDEM resolutions greater than 128 px/deg, the LDEM files have been split by NASA. Although the `MAP_RESOLUTION`, `LINE_SAMPLES` and `LINE_LAST_PIXEL` parameters are identical for each file of one resolution, the `LINE_PROJECTION_OFFSET` and `SAMPLE_PROJECTION_OFFSET` parameters change from file to file. Additional command-line arguments have to be supplied for the specification of the certain LDEM part. The values can be found in the corresponding `*.lbl` file. For these LDEM products, the import script has to be invoked with

```
php ldem_mysql_insert.php additionalFilenamePart LINE_PROJECTION_OFFSET ▼
SAMPLE_PROJECTION_OFFSET
```

where ▼ means that a line break was introduced for typographic reasons, but there must be no line break here. The import command, for example, for the LDEM file `LDEM_1024_00N_15N_330_360.img` would be

```
php ldem_mysql_insert.php 00N_15N_330_360 15359.5 -153600.5
```

according to its label file `LDEM_1024_00N_15N_330_360.LBL`:

```
 1  PDS_VERSION_ID            = "PDS3"
 2
 3  /*** GENERAL DATA DESCRIPTION PARAMETERS ***/
 4  PRODUCT_VERSION_ID        = "V1.05"
 5  DATA_SET_ID               = "LRO-L-LOLA-4-GDR-V1.0"
 6  PRODUCT_ID                = "LDEM_1024_00N_15N_330_360"
 7  INSTRUMENT_HOST_NAME      = "LUNAR RECONNAISSANCE ORBITER"
 8  INSTRUMENT_NAME           = "LUNAR ORBITER LASER ALTIMETER"
 9  INSTRUMENT_ID             = "LOLA"
10  MISSION_PHASE_NAME        = {"COMMISSIONING","NOMINAL MISSION","SCIENCE
11                               MISSION"}
12  TARGET_NAME               = MOON
13  PRODUCT_CREATION_TIME     = 2011-03-15T00:00:00
14  PRODUCER_ID               = LRO_LOLA_TEAM
15  PRODUCER_FULL_NAME        = "DAVID E. SMITH"
16  PRODUCER_INSTITUTION_NAME = "GODDARD SPACE FLIGHT CENTER"
17  OBJECT                    = UNCOMPRESSED_FILE
18  FILE_NAME                 = "LDEM_1024_00N_15N_330_360.IMG"
19  RECORD_TYPE               = FIXED_LENGTH
20  FILE_RECORDS              = 15360
21  RECORD_BYTES              = 61440
22  ^IMAGE                    = "LDEM_1024_00N_15N_330_360.IMG"
23
24    OBJECT                  = IMAGE
25      NAME                  = HEIGHT
26      DESCRIPTION           = "Each sample represents height relative to a
27        reference radius (OFFSET) and is generated using preliminary LOLA data
```

```
                    produced by the LOLA team."
    LINES                     = 15360
    LINE_SAMPLES              = 30720
    SAMPLE_TYPE               = LSB_INTEGER
    SAMPLE_BITS               = 16
    UNIT                      = METER
    SCALING_FACTOR            = 0.5
    OFFSET                    = 1737400.
  END_OBJECT                  = IMAGE
END_OBJECT                    = UNCOMPRESSED_FILE

OBJECT                        = IMAGE_MAP_PROJECTION
 ^DATA_SET_MAP_PROJECTION     = "DSMAP.CAT"
 MAP_PROJECTION_TYPE          = "SIMPLE CYLINDRICAL"
 MAP_RESOLUTION               = 1024 <pix/deg>
 A_AXIS_RADIUS                = 1737.4 <km>
 B_AXIS_RADIUS                = 1737.4 <km>
 C_AXIS_RADIUS                = 1737.4 <km>
 FIRST_STANDARD_PARALLEL      = 'N/A'
 SECOND_STANDARD_PARALLEL     = 'N/A'
 POSITIVE_LONGITUDE_DIRECTION = "EAST"
 CENTER_LATITUDE              = 0. <deg>
 CENTER_LONGITUDE             = 180. <deg>
 REFERENCE_LATITUDE           = 'N/A'
 REFERENCE_LONGITUDE          = 'N/A'
 LINE_FIRST_PIXEL             = 1
 LINE_LAST_PIXEL              = 15360
 SAMPLE_FIRST_PIXEL           = 1
 SAMPLE_LAST_PIXEL            = 30720
 MAP_PROJECTION_ROTATION      = 0.0
 MAP_SCALE                    = 0.0296126469 <km/pix>
 MAXIMUM_LATITUDE             = 15 <deg>
 MINIMUM_LATITUDE             = 0 <deg>
 WESTERNMOST_LONGITUDE        = 330 <deg>
 EASTERNMOST_LONGITUDE        = 360 <deg>
 LINE_PROJECTION_OFFSET       = 15359.5 <pix>
 SAMPLE_PROJECTION_OFFSET     = -153600.5 <pix>
 COORDINATE_SYSTEM_TYPE       = "BODY-FIXED ROTATING"
 COORDINATE_SYSTEM_NAME       = "MEAN EARTH/POLAR AXIS OF DE421"
END_OBJECT                    = IMAGE_MAP_PROJECTION
END
```

Listing 4.2 Example content of a PDS label file. Shown here: The corresponding label file LDEM_1024_00N_15N_330_360.LBL for a LDEM file (.img file), shortened only to give an impression of the PDS format.

4 Creating a Global Lunar Topographic Database

After setting the configuration variables, the script will open a read-only file handle to the specified LRO LDEM file and will establish the database connection (lines 205–216 in the code listing in appendix B.3). Subsequently, the script cycles through the binary LDEM file line by line (lines 218–292). During one cycle, which means one line of the LDEM file has been processed, the script usually forms one SQL query to insert one data row per line sample. Usually, in this context, connotes that the query length will be limited to 24 000 single MySQL insert operations ("inserts") in one SQL query to ensure that one query will not exceed the maxmimum allowed packet size on the MySQL server[15]. This value reflects the maximum number of samples in one LDEM_64 line (23 040 values per line) with an additional margin, since this will be the highest resolution to be used in this thesis. For higher resolutions, the script will automatically split all MySQL insert operations with more than 24 000 inserts (lines 262–274).

The procedure of one cycle shall be illustrated in the following paragraphs. As said before, the number of cycles is determined by the number of lines in the LDEM file (indicated by the LINE_SAMPLES parameter in the PDS label file); the line number $\in [1, \text{LINE\_SAMPLES}]$ is the loop iteration variable $line (line 218). First, the import script reads the entire line of the LDEM file (lines 225 and 226), while the number of bytes to be read is determined by the RECORD_BYTES PDS parameter, starting from

$$\text{start byte} = (\$\text{line} - 1) \cdot \text{RECORD\_BYTES}. \tag{4.1}$$

This line is split up into an array of each single byte, using the unpack function of PHP; the result is stored as an array $line_content (line 226). Each array entry is now one integer value, representing the measured height over MMR for a certain point on the Moon's surface. Furthermore, the preceding part of SQL statement is being prepared in a MySQL command string variable (lines 228–229). The MySQL inserts will simply be appended to this string later.

The script loops through all samples of the line now, beginning with sample 1 and ending with the number of samples per line, as specified by the LINE_SAMPLES PDS parameter. In this loop, the values for all MySQL table columns will be calculated for each sample as well as one ordered 7-tuple is being added to the MySQL command string as one row to be inserted into the corresponding LDEM table. The exact allocation of one height value to a specific point on the Moon's surface for one sample point is calculated using the NASA-supplied equations in

---

[15] The maximum allowed packet size per query is a MySQL configuration option (max_allowed_packet). By default configuration, the max_allowed_packet option is set to 1 MiB. The largest possible size of one single SQL query to a MySQL server is limited to 1 GiB [93].

the LRO data set map projection information for the simple cylindrical projection[16,17]:

$$\text{latitude} = \texttt{CENTER\_LATITUDE} - \frac{\texttt{\$line} - \texttt{LINE\_PROJECTION\_OFFSET} - 1}{\texttt{MAP\_RESOLUTION}} \quad (4.2)$$

$$\text{longitude} = \texttt{CENTER\_LONGITUDE} + \frac{\texttt{\$sample} - \texttt{SAMPLE\_PROJECTION\_OFFSET} - 1}{\texttt{MAP\_RESOLUTION}} \quad (4.3)$$

$$\text{height} = \texttt{\$dn} \cdot \texttt{SCALING\_FACTOR} \quad (4.4)$$

$$\text{planetary radius} = \text{height} + \texttt{OFFSET} \quad (4.5)$$

Out of this information, the spatial coordinates in the Moon ME/PA reference frame can be obtained as follows (cf. section 3.2):

$$x = \text{planetary radius} \cdot \cos\left(\text{latitude} \cdot \frac{\pi}{180°}\right) \cdot \cos\left(\text{longitude} \cdot \frac{\pi}{180°}\right) \quad (4.6)$$

$$y = \text{planetary radius} \cdot \cos\left(\text{latitude} \cdot \frac{\pi}{180°}\right) \cdot \sin\left(\text{longitude} \cdot \frac{\pi}{180°}\right) \quad (4.7)$$

$$z = \text{planetary radius} \cdot \sin\left(\text{longitude} \cdot \frac{\pi}{180°}\right) \quad (4.8)$$

Finally, all these values are combined into a 7-tuple in SQL syntax and appended to the MySQL command string (line 249–258). This cycle repeats for each sample of one LDEM line; at the end, the MySQL command string (a MySQL query) will be executed on the database. After the script execution, all values of the LDEM file have been inserted into the corresponding database table.

---

[16] This information file is distributed as /CATALOG/DSMAP.CAT along with LRO LDEM data. It can be retrieved here: http://imbrium.mit.edu/CATALOG/DSMAP.CAT

[17] All variables written in typewriter font and capital letters are PDS parameters, while all variables written in typewriter font and with a preceding $ are defined variables in the PHP script.

5 Surface Pattern Generation Process

5.1 Anatomy of a Surface Pattern

The existing 3D data, which has been produced by importing and conditioning the lunar digital elevation models (LDEMs) using the technique explained in the previous chapter, is simply a point cloud of discrete measurements of the height of the Moon's surface in relation to its center of mass (COM). Indeed, points do not have a surface or a shape with connected properties (color, reflectance characterics, texture, albedo, and so on) to represent. They are just infinitesimally small markers of locations in space, produced by the sampling of the original Moon's surface. This way, they are only an abstraction of the reality, as the measurement density will have a significant influence on the surface features missed between two successive measurements. The higher the measurement density, ergo the DEM resolution, the higher the model's proximity to reality.

To produce a rendering, a model of the original surface has to be built out of this point cloud, which contains a representation of the discretized[1], real Moon surface. Pursuing the generation of a closed surface out of this point cloud, the points have to somehow be connected with each other, probably new data points need to be inserted by interpolation or existing ones have to be removed.

---

[1] Discretization, in this context, is the procedure of transferring a *continuous* object into individually perceived (*discrete*: separate) pieces, which are called *data points*. The process of discretization always accompanies a loss of information. The compensation of this information loss between two discrete, adjacent data points is called *interpolation*. If additional data points are constructed beyond the boundaries of the sampled data points, this process is called *extrapolation*. Extapolation results are subject to greater uncertainty and often depend on the chosen extrapolation method.

5 Surface Pattern Generation Process

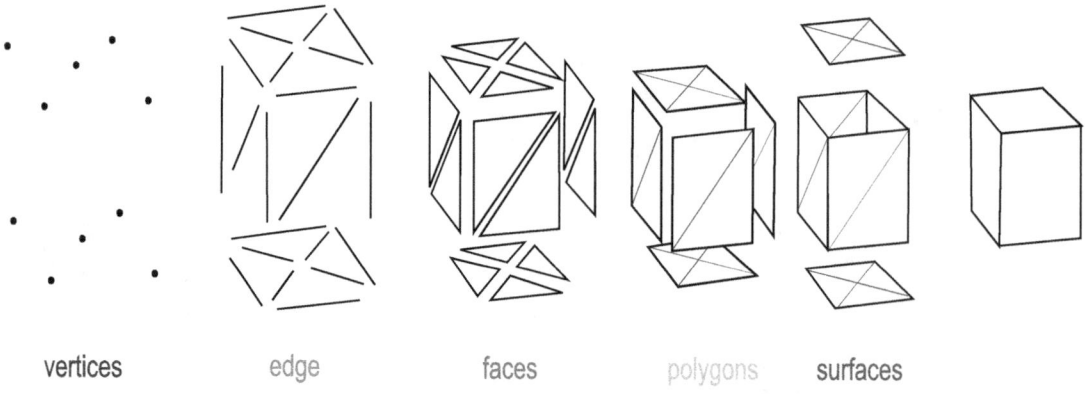

Figure 5.1 Overview of mesh elements in computer graphics. Source: [112], License: Creative Commons Attribution-Share Alike 3.0 Unported.

In the field of computer graphics, this process is known as *mesh modeling*. What has been named a point before, is called a *vertex* (plural: *vertices*; Latin from *vertere*: to turn, revolve [114]). A vertex describes a corner or an intersection of a geometric object, and — in the context of this thesis — one sample point on the real Moon's surface. Vertices can hold other information, such as color or texture, reflectance characteristics and so on. A defined connection between two vertices is called an *edge*, while an area surrounded by closed edges is called a *face*. A set of faces can build a *surface*. A *mesh* stores all of these elements. Picture 5.1 diplays an overview of all mesh elements.

Having said this, a mesh containing one surface (the Moon's surface) has to be produced[2]. There are many ways to connect two or more vertices with each other, resulting in many possible types of polygons. Nevertheless, the used rendering engine, POV-Ray, only supports meshes, which solely consist of triangles [105, pp. 129 f.]. There are a lot of approaches for generating a set of triangles out of a point set; this class of processes is named *point set triangulation*. One common algorithm is the DELAUNAY[3] triangulation.

The DELAUNAY triangulation will produce triangles out of a point set in such a way that no other points of the point set are inside the circumference of any triangle (except the three corner points of the triangle itself). This way, all triangles have the maximum possible interior angle, which is advantageous for computer graphics, since rounding errors are minimized. Figure 5.2 shows the result of a 2D DELAUNAY triangulation for an example point set. [109]

---

[2] To be more exact, multiple meshes will be created. This will be explained later.
[3] Named after Boris Nikolaevich DELAUNAY or Delone (Russian: Бори́с Никола́евич Делоне́; March 15, 1890 – July 17, 1980), Soviet/Russian mathematician. [108]

5.1 Anatomy of a Surface Pattern

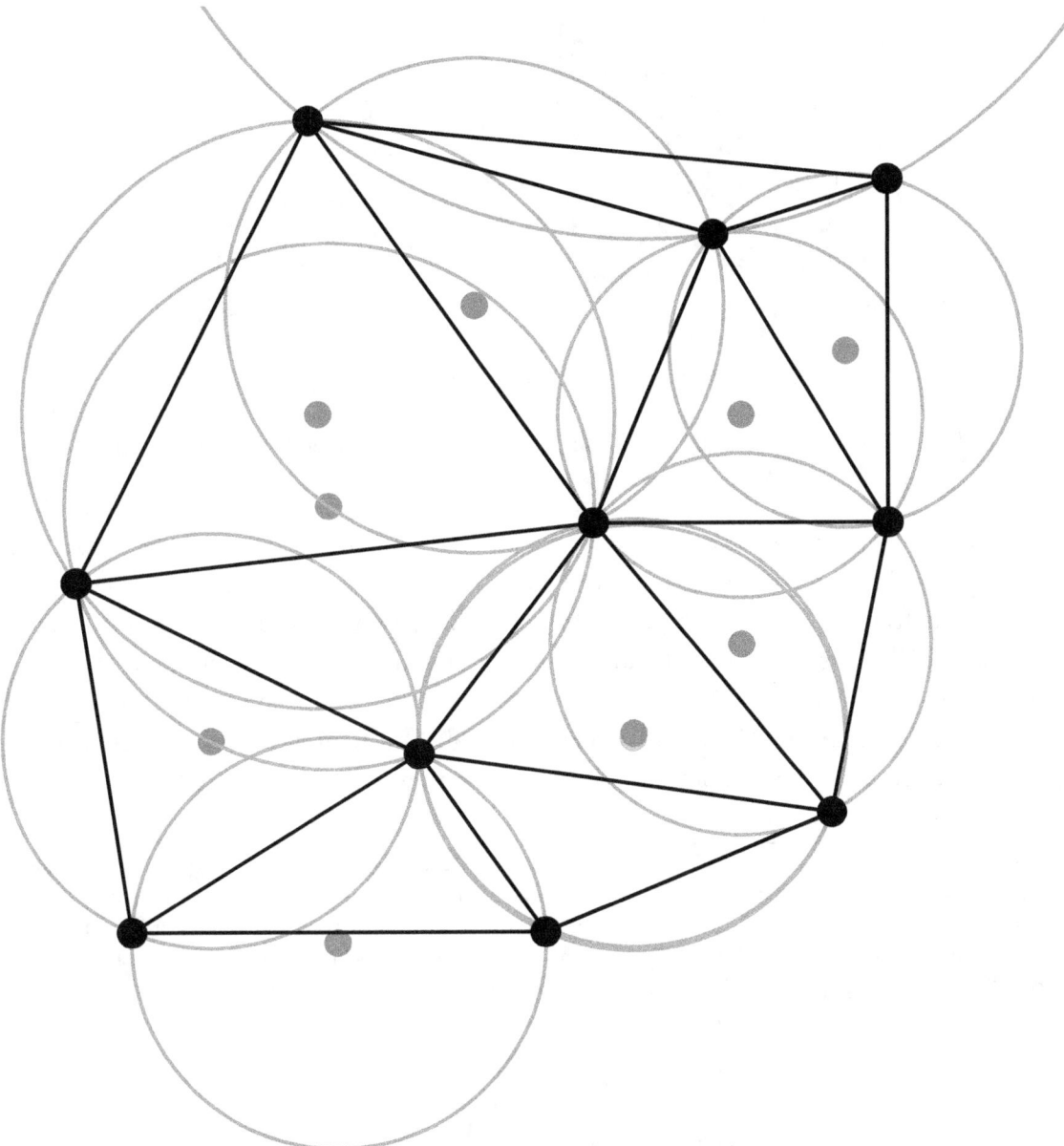

Figure 5.2 Result of a 2D DELAUNAY triangulation for an example point set. The vertices are the black filled dots, and the edges are black lines. The gray circles represent the circumferences of each resulting triangle, while the red filled dots are the center points of each circumference. Vectorized version of [110] by Matthias Kopsch; License: Creative Commons Attribution-Share Alike 3.0 Unported.

5 Surface Pattern Generation Process

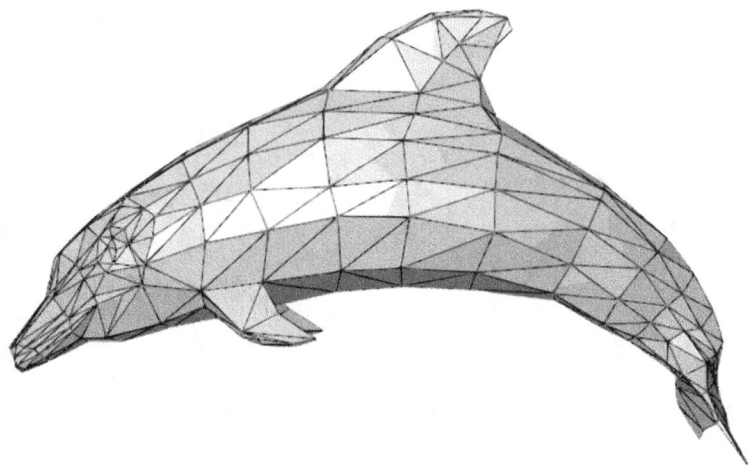

Figure 5.3 A triangulated dolphin. Source: [111]; License: Public Domain.

A DELAUNAY triangulation is also possible in higher dimensions, for example, in 3D space, where it is defined to produce triangles in such a way that no other point of the point set is inside any triangle's circumsphere. However, the 2D DELAUNAY triangulation is sufficient, because LDEM points can be triangulated using their regular grid of selenographic coordinates (see section 3.2). The resulting triangle points are simply interchanged with the corresponding spatial coordinates in the Mean Earth/Polar Axis (ME/PA) reference frame[4] (this is possible because the original selenographic coordinates are kept within the MySQL database).

---

[4] This approach may violate the conditions of a 3D DELAUNAY triangulation, since the 2D regular coordinate grid will be displaced (curved) around a *sphere-like* object in 3D space, meaning that the distances between the points will change. Additionally, there is no direct correspondence between the selenographic coordinates in 2D and the rectangular coordinates in 3D space for this sphere-like object; this is only true for an ideal sphere, since any coordinate in selenographic coordinates can be expressed as

$$\begin{pmatrix} x \\ y \\ z \end{pmatrix} = \begin{pmatrix} r_{\mathrm{C}} \cos \vartheta \cos \varphi \\ r_{\mathrm{C}} \cos \vartheta \sin \varphi \\ r_{\mathrm{C}} \sin \vartheta \end{pmatrix}, \tag{5.1}$$

but for the the Moon's surface, the deviations from this ideal radius r_{C} must be taken into account (this is why a DEM exists). The correspondence between selenographic coordinates and the spatial coordinates in the ME/PA reference frame could be written as

$$\begin{pmatrix} x \\ y \\ z \end{pmatrix} = \begin{pmatrix} (r_{\mathrm{C}} + \delta(\vartheta, \varphi)) \cos \vartheta \cos \varphi \\ (r_{\mathrm{C}} + \delta(\vartheta, \varphi)) \cos \vartheta \sin \varphi \\ (r_{\mathrm{C}} + \delta(\vartheta, \varphi)) \sin \vartheta \end{pmatrix}, \tag{5.2}$$

with the discrete function $\delta(\vartheta, \varphi)$, which gives the measured elevation difference taken from the DEM for a particular point located at (ϑ, φ) on the Moon's surface. But the effect of this consideration is irrelevant for the thesis' result.

After triangulation of the point cloud, the mesh and all triangles in it have to be stored in a POV-Ray file using the *scene description language* (SDL). A POV-Ray file is a simple text file with a defined syntax, which will be parsed by POV-Ray. In the following code snippet, only a segment of a complete POV-Ray file is shown, which is thought to illustrate the syntax to be used for a mesh:

```
mesh {
    triangle {
        <p1_x, p1_y, p1_z>, <p2_x, p2_y, p2_z>, <p3_x, p3_y, p3_z>
        texture { moon }
    }
    triangle {
        <p1_x, p1_y, p1_z>, <p2_x, p2_y, p2_z>, <p3_x, p3_y, p3_z>
        texture { moon }
    }
}
```

This mesh consists of two triangles, which will be defined by three points `<pi_x, pi_y, pi_z>`, where $i \in [1, 2, 3]$ (`pi_x`, `pi_y` and `pi_z` are just placeholders here; these are float values in reality). The line `texture { moon }` inside each triangle object defines the color of this face for POV-Ray (all faces have the same color).

A global POV-Ray mesh covering the entire lunar surface would be extremely huge. For this reason, an alternate approach is used: The Moon's surface is being split into tiles of $(5° + 2\varepsilon) \times (5° + 2\varepsilon)$ in longitude and latitude. This way, 2 592 surface tiles originate for each LDEM resolution. Each of these surface tiles represents one POV-Ray mesh, meaning one surface. Such a surface tile/mesh will be called a *surface pattern* from now on. For a later rendering only those surface patterns will be included in a POV-Ray rendering, which will be visible in this rendering. That means that a later rendering showing a continuous Moon's surface will consist of several overlapping surface patterns. To ensure that no margins are visible between two adjacent surface patterns, the meshes have been created with a margin ε, which is $\varepsilon = 0.5°$ for LDEM resolutions of 4 and 16 px/deg and $\varepsilon = 0.1°$ at 64 px/deg (these values have been chosen arbitrarily).

A surface pattern (the mesh statement) is stored as a single POV-Ray file. POV-Ray's SDL offers the possibility of an include directive (`#include`), which enables the splitting of large POV-Ray files. Thereby all surface patterns can be stored independently from a main POV-Ray file, which is later dynamically generated by the Moon Surface Illumination Simulator (MSIS), and which includes the required surface patterns by referencing their respective files.

5.2 Surface Pattern Generation Process

The surface patterns for a particular LDEM resolution are generated using the generate_pattern.php PHP script, which is available within the /scripts/ directory in the MSISF installation path; a printed version is attached to this thesis as appendix B.4. This script can be run from the command line using the following command:

```
php generate_pattern.php
```

Prior to a script execution, the configuration options inside the script must be correctly set:

```
# ==========================================================
# Configuration of the surface pattern generation process
# ==========================================================

# LDEM resolution (integer) to be used (preconfigured for LDEM_4, LDEM_16 and LDEM_64).
$LDEM = 64;

# Path to pattern repository, where all generated patterns will be placed and where
# all existing patterns are located (w/o trailing slash)
$path_patternDB = "L:/path/to/pattern-repository";

# Path to a temporary directory (w/o trailing slash)
$path_tempdir = "L:/path/to/temp-dir";

# Path to the supplied delaunay2D.exe file (w/o trailing slash)
$path_delaunayHelper = "L:/path/to/MSISF-installation/bin";

# MySQL connection parameters
$mysql_host = "localhost";      # MySQL server, e.g. "localhost" or "localhost:3306"
$mysql_db = "";                 # MySQL database
$mysql_user = "";               # MySQL username for specified database
$mysql_pwd = "";                # MySQL password for specified username
```

The configuration variable $LDEM defines the LDEM resolution for the surface patterns to be generated; the script assumes that a table named LDEM_[resolution], for example, LDEM_64 for the LDEM resolution of 64 px/deg, exists within the specified database (lines 97–101). This table must contain the complete data of an entire NASA LRO LOLA LDEM product, which has been imported using the ldem_mysql_insert.php script (cf. chapter 4) of the MSISF. Additionally, the path to the pattern repository ($path_patternDB), to a temporary working directory ($path_tempdir) and to the delaunay2D.exe file ($path_delaunayHepler), which can be found in the /bin/ directory in the MSISF installation path, must be set.

5.2 Surface Pattern Generation Process

The pattern generation script has been prepared for the LDEM resolutions 4, 16 and 64 px/deg. For the pattern generation of other resolutions, the specification of a margin ε ($off) is necessary:

```
# Surface pattern offset (preconfigured for LDEM_4, LDEM_16 and LDEM_64)
# This offset specifies the latitude and longitude, which will be added to 5°x5° as
# overlap area between the single surface patterns to ensure a closed 3D surface during
# rendering/raytracing. For resolutions greater than 64 px/deg, 0.1° should be sufficient.
# All values of $off are given in degrees latitude/longitude.
switch($LDEM)
{
    case 4:
        $off = 0.5;
        break;
    case 16:
        $off = 0.5;
        break;
    case 64:
        $off = 0.1;
        break;
    default:
        die("No valid LDEM dataset selected.");
        break;
}
```

If all configuration variables are properly set, the script will connect with the specified MySQL database (lines 143–147 of the printed version in appendix B.4). The script will then successively loop through the latitudes and longitudes in steps of 5 degrees (lines 158–321). First, the script constructs the filename of the surface pattern to be generated (line 167); a surface pattern will be named subject to the following scheme:

LDEM_[res]_lat_[lat_start]_[lat_end]_lon_[lon_start]_[lon_end].inc

The script will check if the particular surface pattern exists; an existing surface pattern will not be overwritten at any time, the script will skip the generation of this pattern instead (lines 170–173).

Subsequently (lines 176–216), the pattern generation script will query the database for all points on the Moon's surface within the specified area of $5° \times 5°$ degrees, with a circumferential margin as specified by $off in the configuration section, using a MySQL spatial query (see chapter 4). The return values (latitude, longitude and spatial coordinates in the ME/PA reference frame of each selected point) are stored as a serially indexed array with one entry for each point (lines 218–230). The pair of selenographic coordinates (latitude, longitude) of each selected point will be written into a text file with comma-separated values (called a CSV file), with one line per point (lines 232–234).

A 2D DELAUNAY triangulation is executed on the CSV file, using the `delaunay2D.exe` application (lines 236–273). This application is a compiled MATLAB script, generated using the MATLAB Compiler. The original MATLAB script (the file can be found at `/src/delaunay2D.m`) consists only of a few lines of code:

```
function delaunay2D(csvfile)

    pattern = csvread([csvfile '.csv']);
    triang_pattern = DelaunayTri(pattern);
    dlmwrite([csvfile '_delaunay.csv'], triang_pattern, 'precision', '%li');

end
```

The MATLAB script opens the prepared CSV file and reads the latitude/longitude pairs into its memory. The 2D DELAUNAY triangulation is being offered by MATLAB with the function `DelaunayTri()`. Running the triangulation on the latitude/longitude pairs generates a list of indices; the MATLAB script writes this list into a new CSV file.

The pattern generation script opens this file after the finishing of `delaunay2D.exe`; each line of this CSV file represents a triangle by specifying the line numbers of three latitude/longitude pairs of the first CSV file. For each triangle specified this way, the pattern generation script will create a `triangle` statement in the POV-Ray SDL syntax (lines 275–308). The latitude/longitude pairs are replaced with their corresponding rectangular coordinates in the ME/PA reference frame, as stored in the serially indexed array.

After assembling all `triangle` statements, the surface pattern is written to its pattern file within the pattern repository (lines 310 and 311). After clearing the local loop variables and temporary files (lines 312–321), one loop cycle is complete and the script begins with a the generation of the next surface pattern, if applicable. Once all surface patterns have been generated, the script exits.

5.3 Storage of the Surface Patterns and POV-Ray Mesh Compilation

All surface patterns of the common resolution are stored in a subfolder at the pattern repository path, which is named after the resolution number. The pattern repository itself can be located outside the MSISF installation directory, for example, on a mass storage with sufficient storage size. This way, the MSISF can be installed to the standard Windows program files path, while the larger pattern repository can be placed on a drive with enough memory.

5.3 Storage of the Surface Patterns and POV-Ray Mesh Compilation

The MSIS will later generate a main POV-Ray file, which contains all rendering scene information. This POV-Ray script also includes all needed surface patterns. An example of such a POV-Ray file is the following:

```
#version 3.7;

#declare Orange = rgb <1,0.5,0>;
#declare Red = rgb <1,0,0>;
#declare Yellow = rgb <1,1,0>;
#declare moon = texture
              {
                pigment { color rgb<0.8, 0.8, 0.8> }
                finish
                {
                  ambient 0.0
                  diffuse 0.8
                }
              }

global_settings
{
   charset utf8
   assumed_gamma 1.0
}

camera
{
  perspective
  location <125.340753, 18.2633239, 1793.96436>
  right <0.2277842, -1.5668627, 0.0000365>
  up <0.8009921, 0.1164587, 0.5872385>
  direction <0.4600632, 0.0668672, -0.6407861>
}

light_source
{
  <53424167.691091, 137263136.289498, 3278056.17946521>
  color rgb<1, 1, 1>
  looks_like
  {
    sphere
    {
      0, 1000
      pigment { rgbt 1 }
      hollow
      interior
      {
```

5 Surface Pattern Generation Process

```
44          media
45          {
46            emission 1
47            density
48            {
49              spherical
50              density_map
51              {
52                [0 rgb 0]
53                [60 Orange]
54                [80 Red]
55                [100 Yellow]
56              }
57              scale 1000
58            }
59          }
60        }
61      }
62    }
63  }
64
65  #include "L:\path\to\pattern-repository\64\pattern_LDEM_64_lat_70_75_lon_30_35.inc"
66  #include "L:\path\to\pattern-repository\64\pattern_LDEM_64_lat_75_80_lon_30_35.inc"
67  #include "L:\path\to\pattern-repository\64\pattern_LDEM_64_lat_75_80_lon_25_30.inc"
```

Line 1 declares the POV-Ray version the POV-Ray file is written for. It follows a declaration of the used colors in lines 3–14; here is also the declaration of the surface color for the virtual Moon: It is set to RGB(0.8, 0.8, 0.8), since this value produces reasonable results. If somebody needs to change the color or other appearance settings (reflectance, for example) of all surface patterns, he would only have to edit these lines.

Lines 22–29 implement the camera with their MSIS-calculated geometry, as well as the position of the Sun (the "light source") in lines 31–63. With the lines 65–67, three surface patterns have been included in this example rendering.

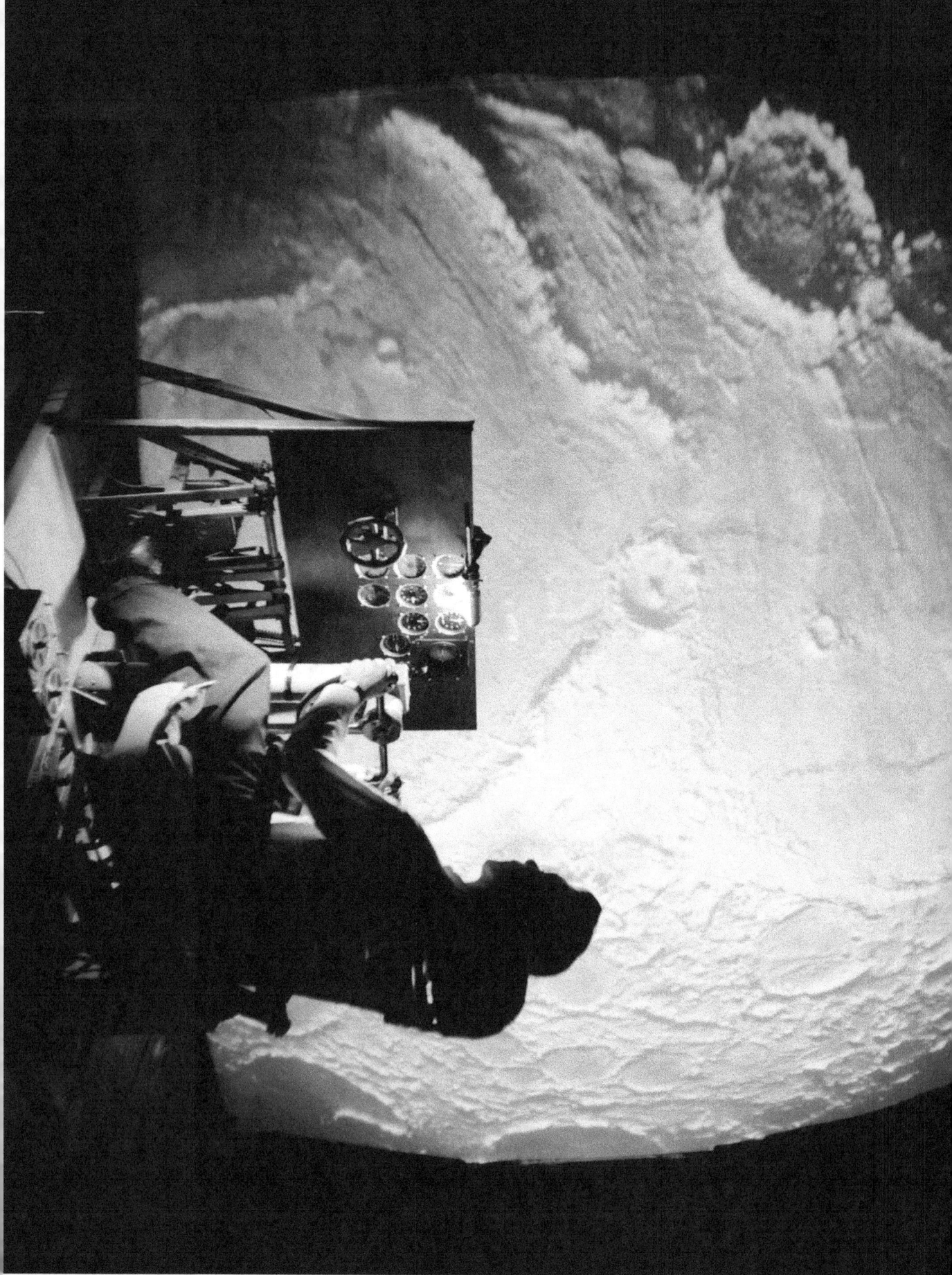

Moon Surface Illumination Simulator (MSIS)

6.1 Software Architecture

The Moon Surface Illumination Simulator (MSIS) is the core application and main user interface of the Moon Illumination Simulation Framework (MSISF). The MSIS is written in C# using Microsoft Visual Studio 2010 Ultimate as integrated development environment (IDE). It is designed using the programming paradigm of object-oriented programming. As such, the MSIS consists of several classes for distinct tasks. The main classes are the `Program`, `Simulation`, `Spacecraft` and `SpacecraftState` class. All other classes are designed as helper classes.

The program flow is controlled by the `Program` class. First, it initiates the parsing and validation process for the user-given command line arguments. After all arguments have been successfully checked, it outputs a configuration overview and initiates the simulation and rendering process. The `Program` class encapsulates all operations within `try-catch` structures and addresses the exceptions thrown by all sub-processes.

The entire simulation process is mapped into the `Simulation` class, which contains and maintains all simulation and rendering parameters, for example, the width and height of the image(s) to be generated, the LDEM resolution to be used, the file system paths to the POV-

---

Chapter Image: *An early "Moon Surface Simulator".* Original image description: Apollo LOLA project. Test subject sitting at the controls: Project LOLA or Lunar Orbit and Landing Approach was a simulator built at Langley to study problems related to landing on the lunar surface. ©1961 NASA Langley Research Center (NASA-LaRC), License: Public Domain. Available at http://mediaarchive.ksc.nasa.gov/detail.cfm?mediaid=41893.

Ray executable, the rendering output directory and the location of the pattern repository. In addition, the `Simulation` class instantiates a new object of the `Spacecraft` class for each given spacecraft state in the batch file or fixed state operation mode. If the operation mode specifying the spacecraft orbit using a set of Keplerian elements or state vectors is used, the `Simulation` class will only instantiate one `Spacecraft` object, but with information regarding its orbit and storing the simulation time(s) to be rendered.

The `Spacecraft` class contains information about the initial location and orientation of the spacecraft, as well as information regarding the spatio-temporal variation of these values. In case of the batch file or fixed state operation modes, an object of the `Spacecraft` class only contains information about the location and orientation of the spacecraft; all other parameters accounting for the variation of the location and orientation are simply set to neutral values.

After all parameters of the `Simulation` class have been properly set, the simulation and rendering process can be initiated by calling the `Simulation.doCalculations()` method. The MSIS will now generate one object of the `SpacecraftState` class per rendering to be done. The `SpacecraftState` class is derived from the `Spacecraft` class and addresses the calculation and application of the spatio-temporal alterations of location and orientation for a given simulation time, based on the initial values given in the properties of a `Spacecraft` object. It will also calculate the Sun's position within the ME/PA reference frame for the given simulation time.

Subsequently, the `Simulation` class calls the Dynamical Surface Pattern Selection Algorithm (DSPSA, see chapter 8), which is implemented itself as a method within the `Simulation` class, for this particular `SpacecraftState` object. The DSPSA determines the surface patterns required to generate the particular rendering; with this information, the `Simulation` class assembles a POV-Ray scene file, containing all information regarding camera location, orientation and geometry as well as the Sun's position for the certain simulation time, surface patterns to be used and some general scene information.

At the same time, the `Simulation` class also generates the XML meta information file (cf. chapter 9), containing general rendering information as well as information regarding the pixel-wise local solar illumination angle. Using the just generated POV-Ray scene file, the MSIS invokes the POV-Ray rendering engine (`pvengine64.exe`) and instructs it to render this file. Additionally, the `Simulation` class initiates the generation of one additional image with an overlay showing a visualization of the rendering meta information (so-called *rendering annotation*), if the user called the MSIS with the command-line option `--rendering-annotation`.

With the completion of all the necessary renderings, the MSIS execution also finishes. During the MSIS execution, the above explained main classes make use of several specialized helper classes. One of these classes is the `tools` class, which contains methods for parsing and evaluating the user-given command-line arguments, time conversion from modified Julian dates

(UTC) to Gregorian dates, some standard math operations (radians to degrees and vice versa, conversion factors, angle normalization, etc.) and an iteration algorithm for solving the KEPLER equation for the eccentric anomaly E.

A second important helper class is the `KeplerOrbit` class, which provides methods for the position determination according to a particular simulation time based on a given set of KEPLERian elements or state vectors. For this purpose, it also contains an algorithm for the conversion of state vectors to KEPLERian elements. All other helper classes implement mathematical constructs or operators like quaternions (`Quaternion` and `RotationQuaternion` class) and vectors (`Vector2D` and `Vector3D` class).

The source code of the MSIS application has been documented in-code; a documentation (available at /doc/MSIS_code_documentation/) has been compiled into the HTML format using doxygen. The following sections will discuss selected passages of the MSIS code, which have a significant importance to the functional principle of the MSIS from a global perspective, or where an implementation seems not to be self-evident. Two algorithms with a substantial amount of theoretical preparatory work, the Dynamical Pattern Selection Algorithm (DSPSA) and the method for determining the local solar illumination angle, have been outsourced into two separate chapters, chapters 8 and 9, respectively.

6.2 Selected Components of the MSIS

6.2.1 Determination of the Sun's Position Using NASA NAIF SPICE

To produce realistically illuminated renderings of the Moon's surface, knowledge about the Sun's position within the Mean Earth/Principal Axis (ME/PA) reference frame for a given simulation time is indispensable. Because an original implementation for finding the Sun's accurate position is very demanding using astrodynamical calculus, the MSIS will use a free and open source, ready-to-use toolkit for astrodynamics, which is called SPICE (abbreviation for *Spacecraft Planet Instrument C-matrix Events*). The SPICE toolkit is offered by NASA's Navigation and Ancillary Information Facility (NAIF), which belongs to the Jet Propulsion Laboratory (JPL) at the California Institute of Technology (Caltech) in the city of Pasadena, California.

The SPICE toolkit is offered in many programming languages, including C/C++, FORTRAN, IDL and MATLAB. Since the MSIS is developed in C#, it seems natural to use the C/C++ version of SPICE, which is also called *CSPICE*. Unfortunately, CSPICE does not directly integrate into a C# application[1]. Because only a few calculations by CSPICE are needed for this thesis, a helper

---

[1] At least a C# wrapper would be needed to use CSPICE in C# directly. Another possibility would be a complete rewrite to C#, but this would be very time-consuming and maintenance intensive.

function for the CSPICE interaction has been written in C++, which is compiled as a dynamic link library (a *.dll file), which can be embedded ("linked") into the C# program. This is only possible because a fixed algorithm with only one input variable (the time) will be used.

CSPICE offers a very simple way for the accurate calculation of the Sun's position within the ME/PA reference frame for a given time. Only a few lines of C++ code lead to the desired result:

```cpp
extern "C" __declspec(dllexport) void __stdcall
calculateSunPosition(char* utc, double* x, double* y, double* z)
{
        furnsh_c("../kernels/moon_pa_de421_1900-2050.bpc");
        furnsh_c("../kernels/moon_080317.tf");
        furnsh_c("../kernels/moon_assoc_me.tf");
        furnsh_c("../kernels/pck00009.tpc");
        furnsh_c("../kernels/naif0009.tls");
        furnsh_c("../kernels/de421.bsp");

        SpiceDouble et;
        utc2et_c(utc, &et);

        SpiceDouble ptarg[3];
        SpiceDouble lt;
        spkpos_c("SUN", et, "MOON_ME", "NONE", "MOON", ptarg, &lt);

        *x = ptarg[0] * 1000;
        *y = ptarg[1] * 1000;
        *z = ptarg[2] * 1000;

        kclear_c();
}
```

Listing 6.1 Implementation of the Sun position calculation using CSPICE in C++.

At the beginning, data about the positions, orientations, sizes and shapes of celestial bodies, reference frames, times, astrodynamical constants, etc. pp. must be available to SPICE. Such data is named *ancillary data* and is either stored in text or binary files, which are called *kernels*. Every kernel has a particular purpose; which kernels to include in an original application depends on the individual case. Depending on their contents, kernels are named SPK (ephemeris), PcK (physical constants, shape and orientation of celestial bodies), FK (reference frames) or LSK (leapseconds) kernels (there are more kernel types, but these are the ones relevant for this thesis). [99, pp. 24 ff.]

Kernels in CSPICE can be loaded using the furnsh_c function. In the context of calculating the Sun's position within the ME/PA reference frame, six kernels have to be loaded.

First, SPICE demands to know about the orientation of the Moon. The binary PcK kernel `moon_pa_de421_1900-2050.bpc` contains high-accuracy lunar orientation data for the years 1900–2050 (line 4 of code listing 6.1). In conjunction with the text FK kernels `moon_080317.tf` and `moon_assoc_me.tf` (lines 5 and 6), which contain information regarding the ME/PA reference frame to be used, SPICE is able to orient the Moon within this frame.

Next, SPICE needs to know about the overall physical constants, orientation, size and shape of the natural bodies of the solar system; information which is loaded into SPICE using the text PcK kernel `pck00009.tpc` (line 7). Additionally, SPICE needs to know about introduced leapseconds for time calculations. This information is stored in the text LSK kernel `naif0009.tls` with the last leapsecond entry at December 31, 2008[2] (line 8). Last, accurate data about positions of the celestial bodies in our solar system is required. Such data is called *ephemerides* and is produced out of complex astrodynamical models. The DE421 ephemerides[3] are one of the most recent ephemerides available; they are stored in the binary SPK kernel `de421.bsp` (line 9).

After loading all these kernels, SPICE is ready to calculate the Sun's position. Prior to this, the time given to the function as ISO 8601 UTC string has to be converted into ephemeris time (ET), because SPICE calculates in ET internally (lines 11–12). After this time conversion, the position of the Sun can be calculated using one single function call (lines 14–16):

```
spkpos_c("SUN", et, "MOON_ME", "NONE", "MOON", ptarg, &lt);
```

The CSPICE `spkpos_c` function returns a position vector of a target body relative to an observing body. The first argument specifies the target body name, the second the time of observation at the observer's location. `MOON_ME` specifies the ME/PA reference frame to be used; `NONE` indicates that no corrections for light time or stellar aberration shall be applied to the result. The fifth argument specifies the observer, which for this thesis is the Moon. The following two arguments specify the output variables; `ptarg` will contain the position of the Sun in the ME/PA reference frame in the unit of km and `lt` the one-way light time from the observer position to the target body (this information is not needed by the MSIS). [98]

---

[2] A leapsecond is a single second, which is added or subtracted to the Coordinated Universal Time (UTC) to adjust the time scale to stay close to mean solar time (realized as Universal Time — UT1). The timing for the introduction of a leapsecond is unpredictable, since a leapsecond compensates the effects of the variation in the Earth's rotation rate. This timing is scheduled by the International Earth Rotation and Reference Systems Service (IERS), which announces a leapsecond if it is foreseeable that UTC will differ more than 0.9 seconds from UT1 any time soon. This way, on one hand, our everyday UTC time scale is sure to differ no more than 0.9 seconds, while other the other hand, an atomic time-precise time unit, the fixed-length SI second, is available.

[3] "DE" stands for *Jet Propulsion Laboratory Development Ephemeris*. They are produced by the Jet Propulsion Laboratory (JPL) in Pasadena, California using complex astrodynamical models of the Solar System. The following number designates the model which has been used. Different DE can cover different ranges of time validity.

Subsequently, the function will store the position of the Sun in the variables x, y and z with their obvious meanings (lines 18–20). Afterwards, the values of these variables will be accessed by the MSIS. Finally, all loaded SPICE kernels will be unloaded and the SPICE session will be cleared using the `kclear_c` function of CSPICE (line 22).

6.2.2 Position Calculation Using a Set of Keplerian Orbit Elements

One essential part of the MSIS operation modes specifying a spacecraft orbit is the calculation of the spacecraft position at a certain simulation time using a set of Keplerian elements. Required for the computation of a position vector $\mathbf{r}(t)$ [AU] and the velocity vector $\dot{\mathbf{r}}(t)$ [$\frac{AU}{d}$], if applicable, is a traditional set of Keplerian elements, which consists of the semi-major axis a [m], eccentricity e [1], argument of periapsis ω [rad], longitude of ascending node (LAN) Ω [rad], inclination i [rad] and the mean anomaly $M_0 = M(t_0)$ [rad] at an epoch t_0 [JD] as well as the considered epoch (simulation time) t [JD], if different from t_0.

The spacecraft's position vector $\mathbf{r}(t)$ and the velocity vector $\dot{\mathbf{r}}(t)$ can then be calculated using the following algorithm[4]:

1. Calculate or set $M(t)$:

    a) If $t = t_0$: $M(t) = M_0$.

    b) If $t \neq t_0$:[5] Determine the time difference Δt in seconds and the mean anomaly $M(t)$ with

    $$\Delta t = 86\,400(t - t_0), \quad M(t) = M_0 + \Delta t \sqrt{\frac{\mu}{a^3}}, \tag{6.1}$$

    whereas $\mu = \mu_{\leftmoon} = 4.902\,801\,076 \cdot 10^{12}\,(\pm 8.1 \cdot 10^4)\,\frac{m^3}{s^2}$ for the Moon as central body. Normalize $M(t)$ to be in $[0, 2\pi)$.

2. Solve Kepler's Equation $M(t) = E(t) - e \sin E$ for the eccentric anomaly $E(t)$ with an appropriate method numerically, e.g. the Newton–Raphson method[6]:

$$f(E) = E - e \sin E - M \tag{6.2}$$

$$E_{j+1} = E_j - \frac{f(E_j)}{\frac{d}{dE_j} f(E_j)} = E_j - \frac{E_j - e \sin E_j - M}{1 - e \cos E_j}, \quad E_0 = M \tag{6.3}$$

---

[4] References: Equations 6.1–6.3 and 6.5–6.7: [4, pp. 22–27]; Equations 6.8 and 6.9: [11, p. 26]; Equation 6.4: [19]; value for $\mu_{\leftmoon}$: [78, p. 305].
[5] Be aware that orbit elements change over time, so be sure to use one set of Orbit Elements given for a certain epoch t_0 only for a small time interval (compared to the rate of changes of the Orbit Elements) around t_0.
[6] Argument (t) omitted for the sake of simplicity.

3. Obtain the true anomaly $\nu(t)$ from

$$\nu(t) = 2 \cdot \operatorname{atan2}\left(\sqrt{1+e}\sin\frac{E(t)}{2}, \sqrt{1-e}\cos\frac{E(t)}{2}\right). \tag{6.4}$$

4. Use the eccentric anomaly $E(t)$ to get the distance to the central body with

$$r_c(t) = a(1 - e\cos E(t)). \tag{6.5}$$

5. Obtain the position and velocity vector $\mathbf{o}(t)$ and $\dot{\mathbf{o}}(t)$, respectively, in the orbital frame (z-axis perpendicular to orbital plane, x-axis pointing to periapsis of the orbit):

$$\mathbf{o}(t) = \begin{pmatrix} o_x(t) \\ o_y(t) \\ o_z(t) \end{pmatrix} = r_c(t) \begin{pmatrix} \cos\nu(t) \\ \sin\nu(t) \\ 0 \end{pmatrix} \tag{6.6}$$

$$\dot{\mathbf{o}}(t) = \begin{pmatrix} \dot{o}_x(t) \\ \dot{o}_y(t) \\ \dot{o}_z(t) \end{pmatrix} = \frac{\sqrt{\mu a}}{r_c(t)} \begin{pmatrix} -\sin E \\ \sqrt{1-e^2}\cos E \\ 0 \end{pmatrix} \tag{6.7}$$

6. Transform $\mathbf{o}(t)$ and $\dot{\mathbf{o}}(t)$ to the inertial frame[7] in bodycentric (in case of the Moon as central body: selenocentric) rectangular coordinates $\mathbf{r}(t)$ and $\dot{\mathbf{r}}(t)$ with the rotation matrices[8] $\mathbf{R}_x(\varphi)$ and $\mathbf{R}_z(\varphi)$ using the transformation sequence

$$\mathbf{r}(t) = \mathbf{R}_z(-\Omega)\mathbf{R}_x(-i)\mathbf{R}_z(-\omega)\mathbf{o}(t) \xrightarrow{o_z(t)=0}$$
$$\begin{pmatrix} o_x(t)(\cos\omega\cos\Omega - \sin\omega\cos i\sin\Omega) - o_y(t)(\sin\omega\cos\Omega + \cos\omega\cos i\sin\Omega) \\ o_x(t)(\cos\omega\sin\Omega - \sin\omega\cos i\cos\Omega) + o_y(t)(\cos\omega\cos i\cos\Omega - \sin\omega\sin\Omega) \\ o_x(t)(\sin\omega\sin i) + o_y(t)(\cos\omega\sin i) \end{pmatrix} \tag{6.8}$$

$$\dot{\mathbf{r}}(t) = \mathbf{R}_z(-\Omega)\mathbf{R}_x(-i)\mathbf{R}_z(-\omega)\dot{\mathbf{o}}(t) \xrightarrow{\dot{o}_z(t)=0}$$
$$\begin{pmatrix} \dot{o}_x(t)(\cos\omega\cos\Omega - \sin\omega\cos i\sin\Omega) - \dot{o}_y(t)(\sin\omega\cos\Omega + \cos\omega\cos i\sin\Omega) \\ \dot{o}_x(t)(\cos\omega\sin\Omega - \sin\omega\cos i\cos\Omega) + \dot{o}_y(t)(\cos\omega\cos i\cos\Omega - \sin\omega\sin\Omega) \\ \dot{o}_x(t)(\sin\omega\sin i) + \dot{o}_y(t)(\cos\omega\sin i) \end{pmatrix}. \tag{6.9}$$

---

[7] With reference to the central body (Moon) and the meaning of i, ω and Ω to its reference frame (the ME/PA reference frame).

[8] $\mathbf{R}_x(\varphi) = \begin{pmatrix} 1 & 0 & 0 \\ 0 & \cos\varphi & -\sin\varphi \\ 0 & \sin\varphi & \cos\varphi \end{pmatrix}, \mathbf{R}_z(\varphi) = \begin{pmatrix} \cos\varphi & -\sin\varphi & 0 \\ \sin\varphi & \cos\varphi & 0 \\ 0 & 0 & 1 \end{pmatrix}$

7. To obtain the position and velocity vector $\mathbf{r}(t)$ and $\dot{\mathbf{r}}(t)$, respectively, in the units AU and AU/d, calculate

$$\mathbf{r}(t)_{[\text{AU}]} = \frac{\mathbf{r}(t)}{1.495\,978\,706\,91 \cdot 10^{11}} \tag{6.10}$$

$$\dot{\mathbf{r}}(t)_{[\text{AU/d}]} = \frac{\dot{\mathbf{r}}(t)}{86\,400 \cdot 1.495\,978\,706\,91 \cdot 10^{11}}. \tag{6.11}$$

This algorithm has been implemented as method `KeplerOrbit.getPosition()` into the MSIS. Since only a position vector is needed as output, the calculations of the velocity vector are cut away. All Keplerian elements, which are input parameters, are properties of the `KeplerOrbit` class.

```java
public Vector3D getPosition(double t)
{
    double M = 0;
    double dt = 0;
    double E = 0;
    double nu = 0;
    double r_c = 0;
    Vector3D o = new Vector3D();
    Vector3D r = new Vector3D();

    if (t == this._epoch)
    {
        M = this._M0;
    }
    else
    {
        dt = 86400 * (t - this._epoch);
        M = this._M0 + dt * Math.Sqrt((_GM) / (Math.Pow(this._a, 3)));
        M = tools.normalizeAngle(M);
    }

    E = tools.solveKeplerForE(M, this._e);
    nu = 2 * Math.Atan2(Math.Sqrt(1 + this._e) * Math.Sin(E / 2), Math.Sqrt(1 - this._e)
            * Math.Cos(E / 2));
    r_c = this._a * (1 - this._e * Math.Cos(E));
    o = r_c * new Vector3D(Math.Cos(nu), Math.Sin(nu), 0);

    r = new Vector3D(
            o.x() * (Math.Cos(this._omega) * Math.Cos(this._Omega) - Math.Sin(this._omega)
            * Math.Cos(this._i) * Math.Sin(this._Omega)) - o.y() * (Math.Sin(this._omega)
            * Math.Cos(this._Omega) + Math.Cos(this._omega) * Math.Cos(this._i)
            * Math.Sin(this._Omega)),
```

```
                    o.x() * (Math.Cos(this._omega) * Math.Sin(this._Omega) - Math.Sin(this._omega)
                    * Math.Cos(this._i) * Math.Cos(this._Omega)) + o.y() * (Math.Cos(this._omega)
                    * Math.Cos(this._i) * Math.Cos(this._Omega) - Math.Sin(this._omega)
                    * Math.Sin(this._Omega)),
                    o.x() * (Math.Sin(this._omega) * Math.Sin(this._i)) + o.y()
                    * (Math.Cos(this._omega) * Math.Sin(this._i)));

        return r;
    }
```

Listing 6.2 Implementation of the spacecraft position calculation using KEPLERian elements into the MSIS.

6.2.3 State Vector Conversion to KEPLERian Elements

Internally, the MSIS will work with KEPLERian elements to calculate the spacecraft position, if an operation mode specifying a spacecraft orbit around the Moon is used. As the MSIS also supports the input of a set of state vectors (position vector $\mathbf{r}(t)$ [m] and velocity vector $\dot{\mathbf{r}}(t)$ [$\frac{m}{s}$]), a conversion to KEPLERian elements is necessary. Given such a set of state vectors, a conversion to KEPLERian elements (semi-major axis a [m], eccentricity e [1], argument of periapsis ω [rad], longitude of ascending node (LAN) Ω [rad], inclination i [rad] and the mean anomaly $M_0 = M(t_0)$ [rad] at the epoch of validity for the given state vectors) can be achieved using the following algorithm[9]:

1. Preparations:

   a) Calculate orbital momentum vector $\mathbf{h}$:

   $$\mathbf{h} = \mathbf{r} \times \dot{\mathbf{r}} \qquad (6.12)$$

   b) Obtain the eccentricity vector $\mathbf{e}$ from

   $$\mathbf{e} = \frac{\dot{\mathbf{r}} \times \mathbf{h}}{\mu} - \frac{\mathbf{r}}{\|\mathbf{r}\|} \qquad (6.13)$$

   with standard gravitational parameter $\mu = \mu_{\mathbb{C}} = 4.902\,801\,076 \cdot 10^{12}\,(\pm 8.1 \cdot 10^4)\,\frac{m^3}{s^2}$ for the Moon as central body.

---

[9] References: Equations 6.12 and 6.22: [4, p. 28]; Eq. 6.13: [15]; Eq. 6.14: [16]; Eq. 6.15 [19]; Eq. 6.16: [14]; Eq. 6.17: [18]; Eq. 6.18: [17]; Eq. 6.19: [16]; Eq. 6.20 [13]; Eq. 6.21: [11, p. 26]; Value for $\mu_{\mathbb{C}}$: [78, p. 305].

c) Determine the vector **n** pointing towards the ascending node and the true anomaly ν with

$$\mathbf{n} = (0,0,1)^T \times \mathbf{h} = (-h_y, h_x, 0)^T \qquad (6.14)$$

$$\nu = \begin{cases} \arccos \frac{\langle \mathbf{e}, \mathbf{r} \rangle}{\|\mathbf{e}\|\|\mathbf{r}\|} & \text{for } \langle \mathbf{r}, \dot{\mathbf{r}} \rangle \geq 0 \\ 2\pi - \arccos \frac{\langle \mathbf{e}, \mathbf{r} \rangle}{\|\mathbf{e}\|\|\mathbf{r}\|} & \text{otherwise.} \end{cases} \qquad (6.15)$$

d) Calculate the eccentric anomaly E:

$$E = \arccos \frac{\|\mathbf{e}\| + \cos \nu}{1 + \|\mathbf{e}\| \cos \nu} \qquad (6.16)$$

2. Calculate the orbit inclination i by using the orbital momentum vector **h**:

$$i = \arccos \frac{h_z}{\|\mathbf{h}\|} \qquad (6.17)$$

(h_z is the third component of **h**).

3. The orbit eccentricity e is simply the magnitude of the eccentricity vector **e**:

$$e = \|\mathbf{e}\| \qquad (6.18)$$

4. Obtain the longitude of the ascending node Ω and the argument of periapsis ω:

$$\Omega = \begin{cases} \arccos \frac{n_x}{\|\mathbf{n}\|} & \text{for } n_y \geq 0 \\ 2\pi - \arccos \frac{n_x}{\|\mathbf{n}\|} & \text{for } n_y < 0 \end{cases} \qquad (6.19)$$

$$\omega = \begin{cases} \arccos \frac{\langle \mathbf{n}, \mathbf{e} \rangle}{\|\mathbf{n}\|\|\mathbf{e}\|} & \text{for } e_z \geq 0 \\ 2\pi - \arccos \frac{\langle \mathbf{n}, \mathbf{e} \rangle}{\|\mathbf{n}\|\|\mathbf{e}\|} & \text{for } e_z < 0 \end{cases} \qquad (6.20)$$

5. Compute the mean anomaly M with help of KEPLER's Equation from the eccentric anomaly E and the eccentricity e:

$$M = E - e \sin E \qquad (6.21)$$

6. Finally, the semi-major axis a is found from the expression

$$a = \frac{1}{\frac{2}{\|\mathbf{r}\|} - \frac{\|\dot{\mathbf{r}}\|^2}{\mu}}. \qquad (6.22)$$

This algorithm has been implemented as an overloaded class constructor of the `KeplerOrbit` class into the MSIS.

```
public KeplerOrbit(Vector3D r, Vector3D dr)
{
    Vector3D h = r % dr;
    Vector3D ev = ((dr % h) / _GM) - (r / r.norm());
    Vector3D n = new Vector3D(0, 0, 1) % h;
    double nu = 0;

    if (r * dr >= 0)
    {
        nu = Math.Acos((ev * r) / (ev.norm() * r.norm()));
    }
    else
    {
        nu = (2*Math.PI - Math.Acos((ev * r) / (ev.norm() * r.norm())));
    }

    double E = Math.Acos((ev.norm() + Math.Cos(nu)) / (1 + (ev.norm() * Math.Cos(nu))));
    double i = Math.Acos(h.z()/h.norm());
    double e = ev.norm();
    double Omega = 0;

    if (n.y() >= 0)
    {
        Omega = Math.Acos(n.x() / n.norm());
    }
    else
    {
        Omega = 2*Math.PI - Math.Acos(n.x() / n.norm());
    }

    double omega = 0;

    if (ev.z() >= 0)
    {
        omega = Math.Acos((n*ev) / (n.norm()*ev.norm()));
    }
    else
    {
        omega = 2*Math.PI - Math.Acos((n * ev) / (n.norm() * ev.norm()));
    }

    double M = E - (e * Math.Sin(E));
    double a = 1 / ((2/r.norm()) - (Math.Pow(dr.norm(),2)/_GM));
```

```
            this.setKeplerElements(a, e, omega, Omega, i, M);
    }
```

Listing 6.3 Implementation of the conversion from state vectors to KEPLERian elements into the MSIS.

6.2.4 Time Calculations

The simulation timepoints are usually given as modified Julian dates (MJD). For the calculation of the Sun's position with SPICE, the declaration of an ISO 8601 UTC string is necessary. Additionally, the MSIS displays dates, which are given as modified Julian dates, as Gregorian dates for verification purposes at the beginning of the rendering process. This should serve to reduce the possible sources of trouble. Jean MEEUS presents an algorithm to convert a given Julian date into a Gregorian date [70, p. 63]. This algorithm has been slightly modified to convert modified Julian dates (MJD) to Gregorian dates, instead of Julian dates (JD); this is actually just a simple addition of the given MJD and 2 400 000.5. Additionally, the algorithm has been customized to return not only the day, but also the time.

The MJD to Gregorian date conversion has been implemented as static method `tools.MJDtoUTC()` into the MSIS[10]:

```
public static DateTime MJDtoUTC(double MJD)
{
    decimal JD = Convert.ToDecimal(MJD) + 2400000.5m;
    decimal T = JD + 0.5m;
    decimal Z = Decimal.Truncate(T);
    decimal F = T - Z;
    decimal A;

    if (Z < 2299161m)
    {
        A = Z;
    }
    else
    {
        decimal alpha = Decimal.Truncate((Z - 1867216.25m) / 36524.25m);
        A = Z + 1 + alpha - Decimal.Truncate(alpha / 4);
    }
```

---

[10] Please note that this code will not address leapseconds, as they are not relevant to the accuracy needed by the MSIS. However, if an accurate conversion to UTC is necessary, one must take the leapseconds into account.

```csharp
        decimal B = A + 1524m;
        decimal C = Decimal.Truncate((B - 122.1m) / 365.25m);
        decimal D = Decimal.Truncate(365.25m * C);
        decimal E = Decimal.Truncate((B - D) / 30.6001m);
        int day = Convert.ToInt32(Decimal.Truncate(B - D - Decimal.Truncate(30.6001m * E) + F));
        int month;
        int year;

        if (E < 14m)
        {
            month = Convert.ToInt32(E - 1m);
        }
        else if (E == 14m || E == 15m)
        {
            month = Convert.ToInt32(E - 13m);
        }
        else
        {
            month = 0;
        }

        if (month > 2m)
        {
            year = Convert.ToInt32(C - 4716m);
        }
        else if (month == 1m || month == 2m)
        {
            year = Convert.ToInt32(C - 4715m);
        }
        else
        {
            year = 0;
        }

        decimal h = F * 24m;
        int hour = Convert.ToInt32(Decimal.Truncate(h));
        decimal m = (h - Decimal.Truncate(h)) * 60m;
        int minute = Convert.ToInt32(Decimal.Truncate(m));
        decimal s = (m - Decimal.Truncate(m)) * 60m;
        int second = Convert.ToInt32(Decimal.Truncate(s));

        return new DateTime(year, month, day, hour, minute, second);
}
```

Listing 6.4 MSIS implementation of the modified Julian date to Gregorian date/time algorithm, based on Jean MEEUS' algorithm [70, p. 63].

6.3 User Interface and MSIS Invocation

6.3.1 General Information

The MSIS is a non-interactive command-line application, which must be called with appropriate arguments. These arguments will specify how the MSIS will operate. In general, there are four distinct operation modes, which can be divided into two separate groups: The first group requires the user to pre-calculate the spacecraft position and orientation, while the second group offers the possibility of specifying an orbit around the Moon. In this case, the actual spacecraft position and orientation will be determined using the temporal evolution of a set of Keplerian elements for each simulation timepoint.

The operation modes "batch file" and "fixed state" do fit in the first group. A consideration of landing trajectories is possible in this mode, as a set of Keplerian elements cannot reflect spacecraft maneuvers (e.g. accelerations/deceleration caused by thrusters), which will likely be a primary simulation circumstance. The second group contains the operation modes "Keplerian elements" and "state vectors", but which are virtually the same, since the MSIS converts given state vectors to a set of Keplerian elements internally.

All possible command-line arguments and their respective meanings are listed in appendix A. The different operation modes are determined by the MSIS by evaluating the user-given command-line arguments. There are one or more mandatory arguments for each operation mode, as well as forbidden and optional ones. Table 6.1 gives an overview of all command-line arguments relating to a certain operation mode. All other listed arguments in appendix A are optional for all operation modes.

Independent from the operation mode used, the MSIS will automatically calculate the position of the Sun within the ME/PA reference frame from the given simulation time. Regardless of how the spacecraft position has been determined (automatically or pre-calculated), the MSIS will set the camera to the given spacecraft position. The camera will be rotated from its default view orientation using the specified rotation quaternion (see chapter 7 for more information regarding the spacecraft/camera orientation), if applicable. The MSIS will always generate a XML meta information file along with a rendering (cf. chapter 9).

A short description of the four operation modes follows; please note the additional information in section 2.3.2.

6.3.2 Batch File Operation Mode

The batch file operation mode is the primary operation mode the MSIS has been designed for. In this mode, a user-specified text file containing a set of so-called *fixed spacecraft states* per one text line will be opened by the application. One line consists of the modified Julian date MJD in

	Kepler Set	State Vectors	Fixed State	Batch File
--attitude (-a)	○	○	✗	✗
--attitude-transition (-at)	○	○	✗	✗
--batch-file	✗	✗	✗	✓
--epoch (-e)	✓	✓	✗	✗
--fixed-state	✗	✗	✓	✗
--kepler-set (-k)	✓	✗	✗	✗
--state-vectors (-s)	✗	✓	✗	✗
--time (-t) --times (-tt) --time-interval	✓	✓	✗	✗

Table 6.1 Overview of the required command-line arguments for the different operation modes of the MSIS. All arguments marked with ✓ are mandatory for the respective operation mode, while all arguments marked with ✗ are not allowed and those with ○ are optional. All additional arguments are optional and have been documented in appendix A.

UTC time, the spacecraft's rectangular position coordinates rx, ry, rz in the ME/PA reference frame and the four components of the rotation quaternion qx, qy, qz and q0, whereas q0 is the real part and qx, qy and qz are the imaginary parts of the quaternion. All values are separated by at least one whitespace or tab:

```
MJD rx ry rz qx qy qz q0
```

All numeric values can be given in decimal or scientific notation (plus sign is optional), even intermixed in one line. An example of a batch file could be:

```
55289 +2.00000E+006 +2.00000E+006 +0 +0.0E+0 0 +3.82683E-001 +9.23879E-001
5.5289E4 -2000000 2000000 0 0 +0.00000E+000 -9.23879E-001 +3.82683E-001
55289 +2.0E6 2E6 0.00 -2.39117e-001 +9.90457E-2 +3.69643e-1 +8.92399E-001
```

This MSIS call will generate one rendering per line. By splitting the input batch file into several pieces, distributed computing of a rendering series is possible, utilizing the power of more than a single machine.

6.3.3 Fixed State Operation Mode

The fixed state operation mode is very similar to the batch file operation mode. Instead of specifying multiple spacecraft states in a separate batch file, the user will specify one spacecraft state directly at the command-line using the `--fixed-state` argument. The fixed state contains the same information as one line of a batch file, but the single values are comma separated and enclosed with curly braces:

```
--fixed-state {MJD,rx,ry,rz,qx,qy,qz,q0}
```

This operation mode is intended for external invoking, for example, out of an external application or script. Additionally, this mode can be used for previewing or rendering a given spacecraft state out of a batch file.

6.3.4 Keplerian Elements Operation Mode

By specifying a set of Keplerian elements, the user enters the Keplerian elements operation mode. The MSIS will automatically calculate the actual spacecraft's position for the given simulation timepoint(s). However, if the user provides no information about the spacecraft orientation using the `--attitude` command-line argument, the spacecraft will point into the direction of nadir. If a user-specified rotation quaternion and multiple simulation timepoints are given, the spacecraft will always look into the same direction for all renderings, unless the user specifies a spacecraft rotation rate using the `--attitude-transition` switch.

This mode assumes a stable orbit to be given by the user.

6.3.5 State Vectors Operation Mode

The usage of state vectors (consisting of one position and one velocity vector) is an alternative way of defining a spacecraft orbit. The MSIS will internally work with a set of Keplerian elements, so a given state vector will be converted. All other information given for the Keplerian elements operation mode also applies for this mode.

6.4 Example Usage

Alteration of the rendering geometry

```
MSIS.exe --fixed-state {55289,2E6,2E6,0,0,0,3.826834323650898E-1,▼
9.238795325112867E-1} --width 1920 --height 1080 --fov 45 --res 16
```

The above MSIS call will produce one rendering in full HD resolution (1920 × 1080 px), using a field of view of 45 degrees and an LDEM resolution of 16 px/deg.

Batch file operation mode with customized POV-Ray path

```
MSIS.exe --batch-file "../input/test-scenarios.tab" --pov-path ▼
"P:\Program Files\POV-Ray\3.7\bin" --batch -f 90 -r 64 -w 1900 -h 1200
```

With these command-line arguments, the MSIS will open the batch file `test-scenarios.tab` and produce as many renderings as the batch file consists of, using a camera field of view of 90 degrees, an LDEM resolution of 64 px/deg and a rendering size of 1900 × 1200 px. Additionally, the user specifies an alternative path to POV-Ray.

Ignoring the Sun's position

```
MSIS.exe --batch-file "../input/test-scenarios.tab" --pov-path ▼
"P:\Program Files\POV-Ray\3.7\bin" --ignore-sun
```

Specifying the `--ignore-sun` command-line switch, the MSIS will ignore the calculated Sun position and will place the light source in the POV-Ray scene to the camera's location. This way, surface parts of the Moon within the field of view will be visible without finding a simulation time at which this particular part of the Moon's surface is illuminated by the Sun. This feature is primarily intended for testing purposes.

Producing a separate rendering with meta information visualization and customized grid spacing for the pixel-wise information

```
MSIS.exe --fixed-state {55289,2E6,2E6,0,0,0,3.826834323650898E-1,▼
9.238795325112867E-1} --width 1920 --height 1080 --res 16 ▼
--gridH 25 --gridV 75 --rendering-annotation
```

The definition of the vertical and horizontal grid spacing for the pixel-wise meta information changes the number and density of pixels, for which pixel-wise information is generated in the XML meta information file (the local solar illumination angle). The `--rendering-annotation` switch instructs the MSIS to generate a separate PNG image, which visualizes the rendering meta information in an image overlay.

Here as he walked by
on the 16th of October 1843
Sir William Rowan Hamilton
in a flash of genius discovered
the fundamental formula for
quaternion multiplication
$i^2 = j^2 = k^2 = ijk = -1$
& cut it on a stone of this bridge

Chapter 7

Spacecraft Orientation and Rotation Model Using Quaternions

Within the MSIS, not only a way to coherently specify a position in space, something achieved by the definition of a reference frame[1], is required, but also a model for the specification of orientations and rotations, for example, for the camera setting in POV-Ray. There exist many ways of performing rotations in $\mathbb{R}^3$. The most often used ones include rotation matrices, EULER angles and quaternions; the latter are used for this thesis.

This chapter provides a short introduction to the nature of quaternions and their application to spatial rotations as well as a short note on the rotation and orientation model to be used within the MSIS.

7.1 Introduction to Quaternions and Spatial Rotation

Here there is no intention to provide an introduction into the complex mathematical theory of quaternions and their algebraic or geometric foundations; instead, interested readers may want to refer to the excellent book of Jack B. KUIPERS [63]. It should be enough to say that quaternions provide an advantageous way to represent spatial rotations in 3-dimensional space. In contrast to EULER angles, they are simpler to compose and compared to rotation matrices they

---

Chapter Image: Quaternion plaque on Brougham (Broom) Bridge, Dublin, Ireland. ©2007 Dr Graeme Taylor (http://straylight.co.uk), University of Bristol. Reproduced with friendly permission.

[1] This thesis uses the Mean Earth/Polar Axis (ME/PA) reference frame, as discussed in section 3.1.

are numerically more stable. Additionally, quaternions avoid the problem of a gimbal lock and they can be stored very efficiently. [67]

Nevertheless, this section shall give a brief note on the usage of quaternions for spatial rotation. Quaternions are a set of numbers that extends the set of complex numbers $\mathbb{C}$ by adding two additional dimensions. Instead of a complex number $c \in \mathbb{C}$, $c = c_0 + \mathrm{i}c_1$ with its real part c_0 and the imaginary part c_1 with the imaginary unit $\mathrm{i} = \sqrt{-1}$, a quaternion

$$q = q_0 + \mathrm{i}q_1 + \mathrm{j}q_2 + \mathrm{k}q_3 \qquad (7.1)$$

with $q \in \mathbb{H}$, $q_0, q_1, q_2, q_3 \in \mathbb{R}$

not only consists of one imaginary part q_1, but of three imaginary parts q_1, q_2 and q_3 with the two additional imaginary units j and k, whereas:

$$\mathrm{i}^2 = \mathrm{j}^2 = k^2 = \mathrm{ijk} = -1 \qquad (7.2)$$
$$\mathrm{ij} = \mathrm{k} = -\mathrm{ji} \qquad (7.3)$$
$$\mathrm{jk} = \mathrm{i} = -\mathrm{kj} \qquad (7.4)$$
$$\mathrm{ki} = \mathrm{j} = -\mathrm{ik} \qquad (7.5)$$

This way, quaternions are a set of 4-dimensional numbers; they are in the class of *hyper-complex numbers*. The set of quaternions is denoted by $\mathbb{H}$ to the honor of their discoverer, Sir William Rowan HAMILTON[2], who first described them in 1843. The real numbers q_0, q_1, q_2 and q_3 are called *components* of the quaternion q. A quaternion can be written as 4-tuple

$$q = (q_0, q_1, q_2, q_3) \qquad (7.6)$$

in $\mathbb{R}^4$. There are several frequently used notations for a quaternion; the above used in equation 7.1 is the most detailed one. However, the author will use the short notation form

$$q = x_0 + \begin{pmatrix} x_1 \\ x_2 \\ x_3 \end{pmatrix} = [x_0, \mathbf{x}], \qquad (7.7)$$

whereas $x_0 \in \mathbb{R}$ represents the *real part* and $\mathbf{x} \in \mathbb{R}^3$, $\mathbf{x} = (x_1, x_2, x_3)^\mathrm{T}$ the 3-dimensional *imaginary part*[3]. Quaternions with a real part of zero are called *pure quaternions*; the set of all

---

[2] Sir William Rowan HAMILTON (August 4, 1805 – September 2, 1865) was an Irish physicist, astronomer, and mathematician. [113]
[3] This is possible, because i, j and k can form an orthonormal base in $\mathbb{R}^3$ using the correspondences $\mathrm{i} \mathrel{\hat=} \mathbf{i} = (1,0,0)^\mathrm{T}$, $\mathrm{j} \mathrel{\hat=} \mathbf{j} = (0,1,0)^\mathrm{T}$ and $\mathrm{k} \mathrel{\hat=} \mathbf{k} = (0,0,1)^\mathrm{T}$, because the cross products of the unit vectors

$$\mathbf{e}_x = (1,0,0)^\mathrm{T}, \ \mathbf{e}_y = (0,1,0)^\mathrm{T}, \ \mathbf{e}_z = (0,0,1)^\mathrm{T} \qquad (7.8)$$

pure quaternions is denoted by $\mathbb{H}_0 \subset \mathbb{H}$. Quaternions do not satisfy the field axioms[4]; they are violating the axiom of commutativity of multiplication (all other axioms are satisfied) [63, p. 6]. Having said this, the quaternion multiplication is *non-commutative*; therefore

$$q_1 q_2 \neq q_2 q_1, \quad q_1, q_2 \in \mathbb{H}. \tag{7.10}$$

The conjugate of quaternion q is denoted by $\bar{q}$ and can be obtained by multiplying the vector part of q with -1:

$$\bar{q} = [q_0, -\mathbf{q}] \tag{7.11}$$

A *quaternion multiplication* of two quaternions q and r, which is — to recall — non-commutative, can be carried out by

$$q \cdot r = [q_0, \mathbf{q}] \cdot [r_0, \mathbf{r}] = [q_0 r_0 - \langle \mathbf{q}, \mathbf{r} \rangle, q_0 \mathbf{r} + r_0 \mathbf{q} + \mathbf{q} \times \mathbf{r}]. \tag{7.12}$$

The *norm* $|q|$ of a quaternion q is computed with

$$|q| = \sqrt{q_0^2 + q_1^2 + q_2^2 + q_3^2}. \tag{7.13}$$

A quaternion with a norm of 1 is called a *unit quaternion*.

There is a 1:1 correspondence between a vector $\mathbf{v} \in \mathbb{R}^3$ and a pure quaternion $v \in \mathbb{H}_0$:

$$\mathbf{v} \leftrightarrow v = 0 + \mathbf{v} \tag{7.14}$$

This means that every vector in 3-dimensional space represents a pure quaternion and every pure quaternion represents a vector in 3-dimensional space. This is important, because it allows a quaternion to operate on vectors of $\mathbb{R}^3$ [63, pp. 114 f.].

Without further proof, rotations in $\mathbb{R}^3$ are possible using a rotation quaternion

$$q_R(\alpha, \mathbf{u}) = \left[\cos \frac{\alpha}{2}, \sin \frac{\alpha}{2} \cdot \mathbf{u} \right], \tag{7.15}$$

whereas α is the rotation angle and $\mathbf{u} \in \mathbb{R}^3$ is the axis of rotation with a norm of 1 ($\|\mathbf{u}\| = 1$, $\mathbf{u}$ is therefore a unit vector) [63, pp. 118 f.]. A rotated point $\mathbf{p}^\star \in \mathbb{R}^3$ originates in the successive

---

of the standard orthonormal base of $\mathbb{R}^3$ give:

$$\mathbf{e}_x \times \mathbf{e}_y = \mathbf{e}_z = -\mathbf{e}_y \times \mathbf{e}_x, \; \mathbf{e}_y \times \mathbf{e}_z = \mathbf{e}_x = -\mathbf{e}_z \times \mathbf{e}_y, \; \mathbf{e}_z \times \mathbf{e}_x = \mathbf{e}_y = -\mathbf{e}_x \times \mathbf{e}_z, \tag{7.9}$$

which can be identified with equations 7.3, 7.4 and 7.5.

[4] In abstract algebra, a field is defined as an algebraic structure with the two operations of addition and multiplication, satisfying the axioms of (1) closure under all operations, (2) associativity and commutativity of all operations, (3) existence of neutral elements and inverses for addition and subtraction and (4) distributivity of multiplication over addition. The operations of subtraction and division are implicitly defined as inverse operations of addition and multiplication, respectively. [63, pp. 5 f.]

quaternion multiplication of the rotation quaternion $q_R(\alpha, \mathbf{u})$, the quaternion p of its original point $\mathbf{p}$ and the conjugate $\bar{q}_R(\alpha, \mathbf{u})$ of the rotation quaternion:

$$p^\star = [0, \mathbf{p}^\star] = q_R(\alpha, \mathbf{u}) \cdot [0, \mathbf{p}] \cdot \bar{q}_R(\alpha, \mathbf{u}) \tag{7.16}$$

That means such a quaternion multiplication would always have to result in a pure quaternion, which represents a (new) vector in 3-dimensional space. [63, pp. 119 f.]

Theorem 7.1. A successive quaternion multiplication of a rotation quaternion $q_R = [s, t\mathbf{u}]$, whereas $s = \cos\frac{\alpha}{2}$ and $t = \sin\frac{\alpha}{2}$, a pure quaternion x and the conjugate $\bar{q}_R$ of the rotation quaternion will always yield a pure quaternion.

Proof.

$$[s, t\mathbf{u}] \cdot [0, \mathbf{x}] \cdot [s, -t\mathbf{u}]$$
$$= [-\langle t\mathbf{u}, \mathbf{x}\rangle, s\mathbf{x} + t\mathbf{u} \times \mathbf{x}] \cdot [s, -t\mathbf{u}]$$
$$= [-t\langle \mathbf{u}, \mathbf{x}\rangle, s\mathbf{x} + t\mathbf{u} \times \mathbf{x}] \cdot [s, -t\mathbf{u}]$$
$$= [-t\langle \mathbf{u}, \mathbf{x}\rangle s - \langle s\mathbf{x} + t\mathbf{u} \times \mathbf{x}, -t\mathbf{u}\rangle,$$
$$\quad -t\langle \mathbf{u}, \mathbf{x}\rangle \cdot (-t\mathbf{u}) + s(s\mathbf{x} + t\mathbf{u} \times \mathbf{x}) + (s\mathbf{x} + t\mathbf{u} \times \mathbf{x}) \times (-t\mathbf{u})]$$
$$= [-st\langle \mathbf{u}, \mathbf{x}\rangle - \langle s\mathbf{x} + t\mathbf{u} \times \mathbf{x}, -t\mathbf{u}\rangle,$$
$$\quad t^2\langle \mathbf{u}, \mathbf{x}\rangle\mathbf{u} + s(s\mathbf{x} + t\mathbf{u} \times \mathbf{x}) + (s\mathbf{x} + t\mathbf{u} \times \mathbf{x}) \times (-t\mathbf{u})]$$
$$= [-st\langle \mathbf{u}, \mathbf{x}\rangle - (\langle s\mathbf{x}, -t\mathbf{u}\rangle + \langle t\mathbf{u} \times \mathbf{x}, -t\mathbf{u}\rangle),$$
$$\quad t^2\langle \mathbf{u}, \mathbf{x}\rangle\mathbf{u} + s(s\mathbf{x} + t\mathbf{u} \times \mathbf{x}) + (s\mathbf{x} + t\mathbf{u} \times \mathbf{x}) \times (-t\mathbf{u})]$$
$$= [-st\langle \mathbf{u}, \mathbf{x}\rangle - (-st\langle \mathbf{x}, \mathbf{u}\rangle - t^2\langle \mathbf{u} \times \mathbf{x}, \mathbf{u}\rangle),$$
$$\quad t^2\langle \mathbf{u}, \mathbf{x}\rangle\mathbf{u} + s(s\mathbf{x} + t\mathbf{u} \times \mathbf{x}) + (s\mathbf{x} + t\mathbf{u} \times \mathbf{x}) \times (-t\mathbf{u})]$$
$$= \Big[\underbrace{-st\langle \mathbf{u}, \mathbf{x}\rangle + st\langle \mathbf{x}, \mathbf{u}\rangle}_{=0} + t^2\underbrace{\langle \mathbf{u} \times \mathbf{x}, \mathbf{u}\rangle}_{=\det(\mathbf{u}, \mathbf{x}, \mathbf{u})=0},$$
$$\quad t^2\langle \mathbf{u}, \mathbf{x}\rangle\mathbf{u} + s(s\mathbf{x} + t\mathbf{u} \times \mathbf{x}) + (s\mathbf{x} + t\mathbf{u} \times \mathbf{x}) \times (-t\mathbf{u})\Big]$$
$$= [0, t^2\langle \mathbf{u}, \mathbf{x}\rangle\mathbf{u} + s(s\mathbf{x} + t\mathbf{u} \times \mathbf{x}) + (s\mathbf{x} + t\mathbf{u} \times \mathbf{x}) \times (-t\mathbf{u})]$$
$$= [0, t^2\langle \mathbf{u}, \mathbf{x}\rangle\mathbf{u} + s^2\mathbf{x} + 2st(\mathbf{u} \times \mathbf{x}) - t^2(\mathbf{u} \times \mathbf{x}) \times \mathbf{u}]$$

∎

7.2 Spacecraft Orientation and Rotation Model

This section explains how the virtual spacecraft/camera will be rotated, if a rotation quaternion q_o has been given by the MSIS user. If no rotation quaternion is given, the camera will always point into the direction of nadir and the following text is not applicable.

If a user-given rotation quaternion q_o has been set, the virtual spacecraft/camera is initially aligned into the direction $\mathbf{d} = (-1, 0, 0)^\mathrm{T}$ from its actual position; the camera direction vector $\mathbf{c}_\text{direction}$ (see section 8.2) will be a scalar multiple of

$$\mathbf{c}_\text{pos} + \mathbf{d} = \begin{pmatrix} c_{\text{pos},x} + d_x \\ c_{\text{pos},y} + d_y \\ c_{\text{pos},z} + d_z \end{pmatrix}, \tag{7.17}$$

while the camera's right vector $\mathbf{c}_\text{right}$ will always point into the direction of $\mathbf{r} = (0, -\frac{w}{h}, 0)^\mathrm{T}$, whereas w is the rendering image width and h the height.

The user-given rotation quaternion q_o will be applied to both direction vectors d and r using quaternion multiplication, resulting in two rotated vectors $\mathbf{d}^\star$ and $\mathbf{r}^\star$:

$$[0, \mathbf{d}^\star] = q_o \cdot [0, \mathbf{d}] \cdot \bar{q}_o \tag{7.18}$$

$$[0, \mathbf{r}^\star] = q_o \cdot [0, \mathbf{r}] \cdot \bar{q}_o \tag{7.19}$$

It is assumed that the user-given quaternion q_o is a valid rotation quaternion, especially that $|q_o| = 1$. For the rotation, the axis definition of the ME/PA reference frame applies, that means the rotation will be carried out in a right-handed, cartesian coordinate system.

Dynamical Surface Pattern Selection

A generated POV-Ray surface mesh for the entire Moon can consume a lot of space in memory, depending on the chosen LOLA LDEM resolution, so that a rendering with POV-Ray may be impossible at a certain resolution. The finest resolution computed for this thesis was 64 px/deg. The generated POV-Ray pattern repository for this LOLA LDEM resolution has a file size over 115 GiB. To render the full Moon surface at this resolution, all data has to be parsed by POV-Ray (and to be stored in the system memory). Such a rendering would be impossible on a standard computer.

So an algorithm is used by the MSISF to dynamically select only those patterns out of the pattern repository, which will be visible later in the rendered image. This algorithm has been named *Dynamical Surface Pattern Selection Algorithm* (DSPSA). The DSPSA is based on the ray tracing technique: A ray will be shot from the camera's focal point for each pixel on the image plane and it will be examined which points on the Moon's surface encounter the ray. The needed surface patterns can be identified this way. This technique needs some preliminary considerations and preparations before implementation.

First, a brief note on the mathematical technique used for ray tracing on the example of a sphere shall be given in section 8.1. This section is followed by some information on the internal camera geometry to be used in POV-Ray. With the help of both previous sections, this chapter finally closes with a discussion of the implementation of the surface pattern selection, explaining the DSPSA in detail.

Chapter Image: *This is how surface patterns would have been selected manually, without the DSPSA.* Original image description: Apollo Project. Artists used paintbrushes and airbrushes to recreate the lunar surface on each of the four models comprising the LOLA simulator. ©1964 NASA/Langley Research Center (LaRC), License: Public Domain. Available at http://www.nasaimages.org/luna/servlet/detail/nasaNAS~2~2~6659~108281.

8 Dynamical Surface Pattern Selection

8.1 Ray Tracing with a Sphere

One of the most frequently used examples for the introduction of the ray tracing algorithm seems to be a scene with a sphere [24, pp. 56–59][22, chapter 2][23, pp. 7, 8, 116–119]. Fortunately, this kind of scene is needed for the purpose of the DSPSA.

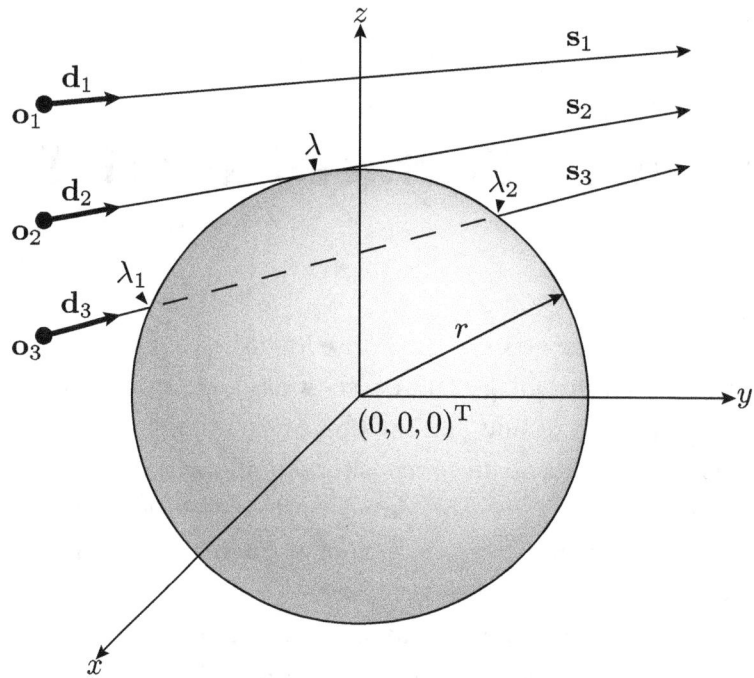

Figure 8.1 Example ray tracing scene with a sphere. The sphere's center is placed at the origin $(0,0,0)^T$ of the right-handed coordinate system. Three example rays s_1, s_2 and s_3 are shot: s_1 misses the sphere, s_2 is tangential to the sphere's surface and s_3 intersects the sphere twice. In ray 2 and 3, the intersection points can be calculated solving equation 8.4 for λ or λ_1 and λ_2, respectively, and substituting the parameter(s) into formula 8.2.

At first, a sphere with radius r, whose center is located at the origin $(0,0,0)^T$ of a cartesian coordinate system, is considered. All points $\mathbf{p}_i$ on the surface obviously satisfy the condition $\|\mathbf{p}_i\| = r = \sqrt{p_{i,x}^2 + p_{i,y}^2 + p_{i,z}^2}$ and therefore the power of $\|\mathbf{p}_i\|$ by 2 will be

$$\|\mathbf{p}_i\|^2 = r^2 = p_{i,x}^2 + p_{i,y}^2 + p_{i,z}^2 \\ = \langle \mathbf{p}_i, \mathbf{p}_i \rangle. \tag{8.1}$$

A ray $\mathbf{s}(\lambda)$ is now shot from its origin $\mathbf{o}$ into the direction $\mathbf{d}$. This ray can be written as

$$\mathbf{s}(\lambda) = \mathbf{o} + \lambda\,\mathbf{d}, \tag{8.2}$$

where $\lambda \in [0, \infty)$ is a measure for the distance between the actual photon position $\mathbf{s}$ and its origin $\mathbf{o}$. Due to this equation, all points on the ray $\mathbf{s}$ are known. The question now is whether the ray encounters a point on the sphere's surface or not. If a surface point $\mathbf{p}_i$ lies on the ray, $\mathbf{p}_i$ must be $\mathbf{p}_i = \mathbf{s}(\lambda)$ for one λ:

$$\begin{aligned} r^2 = \langle \mathbf{p}_i, \mathbf{p}_i \rangle = \langle \mathbf{s}(\lambda), \mathbf{s}(\lambda) \rangle &= \langle \mathbf{o} + \lambda\mathbf{d}, \mathbf{o} + \lambda\mathbf{d} \rangle \\ &= \langle \mathbf{o}, \mathbf{o} \rangle + 2\langle \mathbf{o}, \mathbf{d} \rangle \lambda + \langle \mathbf{d}, \mathbf{d} \rangle \lambda^2 \end{aligned} \tag{8.3}$$

After subtracting r^2 from equation 8.3, a quadratic equation emerges. The two solutions of the quadratic equation for λ are given by the quadratic formula

$$\lambda = \frac{-2\langle \mathbf{o}, \mathbf{d} \rangle \pm \sqrt{4\langle \mathbf{o}, \mathbf{d} \rangle^2 - 4\langle \mathbf{d}, \mathbf{d} \rangle \langle \mathbf{o}, \mathbf{o} \rangle + 4\langle \mathbf{d}, \mathbf{d} \rangle r^2}}{2\langle \mathbf{d}, \mathbf{d} \rangle}. \tag{8.4}$$

Based on the discriminant $D = 4\langle \mathbf{o}, \mathbf{d} \rangle^2 - 4\langle \mathbf{d}, \mathbf{d} \rangle \langle \mathbf{o}, \mathbf{o} \rangle + 4\langle \mathbf{d}, \mathbf{d} \rangle r^2$, a decision can be made whether the ray will encounter the sphere or not. If $D < 0$ there will be no solutions in the set of the real numbers, since the square root of a negative number is not defined there. In this case, the ray will not encounter the sphere. In the case $D = 0$ there will be only one solution; the ray is tangential to the sphere. Finally, the ray intersects the sphere twice, if $D > 0$. The hit surface point(s) can be obtained by evaluating equation 8.2 with the calculated parameter(s) λ.

8.2 Camera Geometry

To make a prediction concerning which parts of the Moon's surface will be visible, not only one ray must be shot, but one ray for each pixel of the rendered image. This means 1 048 576 rays must be shot and traced for a rendering size of 1 024 px × 1 024 px. For the correct origin and direction declaration of each single ray, a consideration of the camera geometry of POV-Ray is necessary. The MSIS uses the model of a standard pinhole camera, which is a perspective camera type in POV-Ray. Figure 8.2 shows the camera geometry, as it is set in POV-Ray by the MSIS.

The camera position $\mathbf{c}_{\text{pos}}$ is set to the actual spacecraft location — no consideration of the spacecraft and instrument geometry will be made by the MSIS. This means spacecraft and

8 Dynamical Surface Pattern Selection

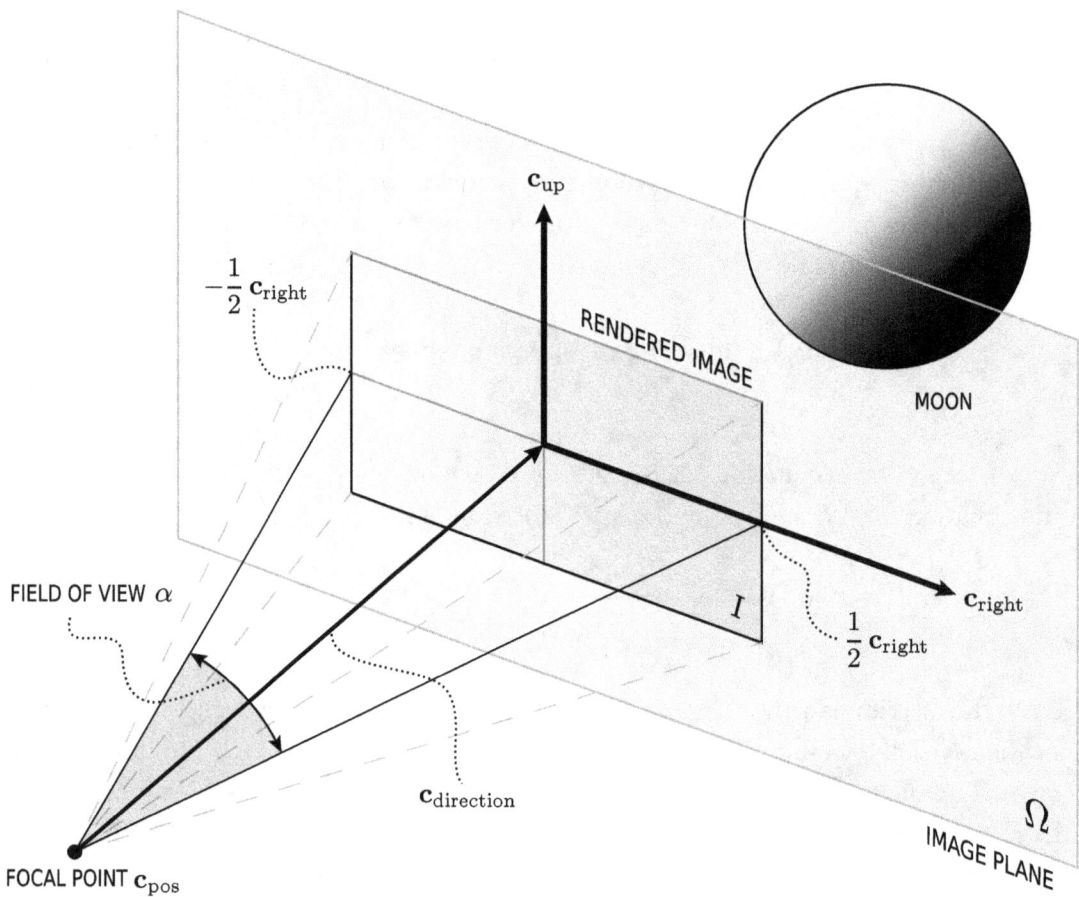

Figure 8.2 POV-Ray camera geometry to be used. The focal point c_{pos} of the camera is identical with the spacecraft/camera location; c_{pos} is a spatial coordinate within the ME/PA reference frame. The user-given field of view angle α ordains the length of the camera direction vector $c_{direction}$ and, by implication, the distance between the focal point c_{pos} and the image plane Ω.

The pixels of the later rendered picture are a strict subset of all points on the image plane Ω, which is defined by the camera-up vector c_{up} and the camera-right vector c_{right}. By definition, the rendered image is an axes-parallel rectangle with a width of $\|c_{right}\|$ and a height of $\|c_{up}\|$, while the axes, in this case, are c_{right} and c_{up} itself; the rectangle is centered at $c_{pos} + c_{direction}$.

c_{right} and c_{up} are influenced by the user-specified pixel width and height of the image to be rendered. The determination of the camera orientation is explained in chapter 7.

camera instrument will be one single spatial point[1]. This way, the terms *spacecraft* and *camera* will be used interchangeably henceforth.

The camera direction vector $\mathbf{c}_{\text{direction}}$ defines in which direction the camera will be pointed starting from the camera location $\mathbf{c}_{\text{pos}}$. It will be calculated by involving several steps. First, the direction $\mathbf{a}$ of the vector is determined by the quaternion rotation of the vector $(-1, 0, 0)^{\mathrm{T}}$ (this is the default camera direction, c.f. chapter 7) with the orientation quaternion $q_o \in \mathbb{H}$ if a user-given camera orientation will be used[2]:

$$[0, \mathbf{a}] = q_o \cdot [0, (-1, 0, 0)^{\mathrm{T}}] \cdot \overline{q}_o \tag{8.5}$$

If no user-supplied camera orientation is given, the camera will be nadir-looking:

$$\mathbf{a} = -\mathbf{c}_{\text{pos}} \tag{8.6}$$

Additionally, the length of the camera direction vector $\mathbf{c}_{\text{direction}}$ defines the focal length of the pinhole camera in POV-Ray; $\mathbf{c}_{\text{direction}}$ has to satisfy the equation

$$\|\mathbf{c}_{\text{direction}}\| = \frac{\frac{1}{2}\|\mathbf{c}_{\text{right}}\|}{\tan\frac{\alpha}{2}}, \tag{8.7}$$

where

$$[0, \mathbf{c}_{\text{right}}] \stackrel{\text{def.}}{=\!=} q_o \cdot \left[0, \left(0, -\frac{w}{h}, 0\right)^{\mathrm{T}}\right] \cdot \overline{q}_o \tag{8.8}$$

is a POV-Ray specific vector to set the ratio of the image plane, α is the field of view (default 40 degrees) and w and h are the width and height of the rendered image in pixels, respectively. For a default rendering width and height of $w = 1\,024\,\text{px}$ and $h = 1\,024\,\text{px}$ and the default field of view $\alpha = 40°$ the length of the camera pointing vector $\mathbf{c}_{\text{direction}}$ becomes

$$\|\mathbf{c}_{\text{direction}}\| = \frac{\frac{1}{2}\|(0, -1, 0)^{\mathrm{T}}\|}{\tan\frac{40°}{2}} \approx 1.373\,74. \tag{8.9}$$

The camera direction vector $\mathbf{c}_{\text{direction}}$ itself can now be calculated with

$$\mathbf{c}_{\text{direction}} = \frac{\mathbf{a}}{\|\mathbf{a}\|} \cdot \|\mathbf{c}_{\text{direction}}\| = \frac{\mathbf{a}}{\|\mathbf{a}\|} \cdot \frac{\frac{1}{2}\|\mathbf{c}_{\text{right}}\|}{\tan\frac{\alpha}{2}}. \tag{8.10}$$

---

[1] In fact, spacecraft and camera instrument are the same for the MSIS. If another instrument pointing is to be computed, external calculations with respective translations/rotations relative to the camera position and orientation have to be done.

[2] At default, the camera points in the direction of nadir $\mathbf{p}_{\mathrm{C}} = (0, 0, 0)^{\mathrm{T}}$, rendering an image perpendicular to the Moon's surface at the sub-satellite point. The necessary orientation quaternion will be calculated by the MSIS, unless the user specifies the camera orientation manually; the camera orientation might be set by the user using a rotation quaternion and an orientation transition vector or by a fixed orientation quaternion at a given simulation time. Details of the spacecraft orientation notation and rotation model have been discussed previously in chapter 7. From this chapter, the spacecraft orientation quaternion $q_o \in \mathbb{H}$ will be used in this chapter.

8 Dynamical Surface Pattern Selection

The camera-up vector $\mathbf{c}_{\text{up}}$ is defined to be perpendicular and normal to the plane defined by the camera-right vector $\mathbf{c}_{\text{right}}$ and the camera direction vector $\mathbf{c}_{\text{direction}}$. Hence, the camera-up vector $\mathbf{c}_{\text{up}}$ can be obtained by calculating the cross product of both vectors:

$$\mathbf{c}_{\text{up}} = \frac{\mathbf{c}_{\text{right}} \times \mathbf{c}_{\text{direction}}}{\|\mathbf{c}_{\text{right}} \times \mathbf{c}_{\text{direction}}\|} \tag{8.11}$$

The camera-up vector $\mathbf{c}_{\text{up}}$ and the camera-right vector $\mathbf{c}_{\text{right}}$ now define the plane, on which the scene will be projected. The later image, which is generated by the rendering process, is a convex, bounded and closed set $I = \{(x,y) \mid x \in \{1,2,\ldots,w\} \subset \mathbb{N}^+, y \in \{1,2,\ldots,h\} \subset \mathbb{N}^+\}$ of $w \cdot h$ discrete pixels as ordered pairs (x,y) on the image plane Ω. The center of pixel $(1,1)$ will lie at $\mathbf{c}_{\text{pos}} + \mathbf{c}_{\text{direction}} + \frac{h-1}{2h}\mathbf{c}_{\text{up}} + \frac{1-w}{2w}\mathbf{c}_{\text{right}}$. The position of the center of one arbitrary pixel (x,y) can be written as

$$\text{PixelPos}: \{1,2,\ldots,w\} \subset \mathbb{N}^+ \times \{1,2,\ldots,h\} \subset \mathbb{N}^+ \to \mathbb{R}^3, (x,y) \mapsto$$
$$\mathbf{c}_{\text{pos}} + \mathbf{c}_{\text{direction}} + \frac{1}{2}\mathbf{c}_{\text{up}} - \frac{1}{2}\mathbf{c}_{\text{right}} + \frac{x-1}{w}\mathbf{c}_{\text{right}} - \frac{y-1}{h}\mathbf{c}_{\text{up}}$$
$$+ \frac{1}{2w}\mathbf{c}_{\text{right}} - \frac{1}{2h}\mathbf{c}_{\text{up}}$$
$$= \mathbf{c}_{\text{pos}} + \mathbf{c}_{\text{direction}} + \frac{1+h-2y}{2h}\mathbf{c}_{\text{up}} - \frac{1+w-2x}{2w}\mathbf{c}_{\text{right}} \tag{8.12}$$

and therefore the image plane Ω as

$$\Omega: \mathbb{R} \times \mathbb{R} \to \mathbb{R}^3, (s_x, s_y) \mapsto \mathbf{c}_{\text{pos}} + \mathbf{c}_{\text{direction}} + \frac{1+h-2s_y}{2h}\mathbf{c}_{\text{up}} - \frac{1+w-2s_x}{2w}\mathbf{c}_{\text{right}}. \tag{8.13}$$

This way, the parameters s_x and s_y of the image plane Ω are a strict superset of the later rendered image's pixels. These two parameters are called *rendering coordinates* (s_x, s_y). They differ from the pixels (x,y) of the later rendered images[3] as they can be non-integer, negative and/or numbers not in the defined range (width and height of the later rendered image):

$$I = \{(x,y) \mid x \in \{1,2,\ldots,w\}, y \in \{1,2,\ldots,h\}\} \subset \Omega = \{(s_x, s_y) \mid s_x, s_y \in \mathbb{R}\} \tag{8.14}$$

With reference to section 8.1, the direction vector $\mathbf{d}(x,y)$ for one ray starting at $\mathbf{o} = \mathbf{c}_{\text{pos}}$ and intersecting a certain pixel on the image plane Ω can now be determined using

$$\begin{aligned}\mathbf{d}(x,y) &= \mathbf{c}_{\text{direction}} + \frac{1+h-2y}{2h}\mathbf{c}_{\text{up}} - \frac{1+w-2x}{2w}\mathbf{c}_{\text{right}} \\ &= \frac{\mathbf{a}}{\|\mathbf{a}\|} \cdot \frac{\frac{1}{2}\|\mathbf{c}_{\text{right}}\|}{\tan\frac{\alpha}{2}} + \frac{1+h-2y}{2h}\mathbf{c}_{\text{up}} - \frac{1+w-2x}{2w}\mathbf{c}_{\text{right}}.\end{aligned} \tag{8.15}$$

---

[3] Generally, a digital image consists of discrete pixels $\in \mathbb{N}^+$. Since the plane equation for Ω shall yield continuous values in $\mathbb{R}^3$ by definition, its two parameters need to be in the set of real numbers.

8.3 MSIS Implementation

The DSPSA is implemented as method `DSPSA(Vector3D rayDirection, Vector3D camPos, ArrayList patternList)` into the `Simulation` class. The method expects three arguments: As required for the ray tracing, the direction vector of one ray shot (`rayDirection`) must be given, as well as its origin (`camPos`). Furthermore, the third argument `patternList` is an array of strings, containing the file names of all required surface patterns for this particular rendering so far.

The algorithm checks whether a ray shot from a specific location into a defined direction hits the Moon's surface or not[4]. If so, the DSPSA will determine which particular part of the Moon's surface has been hit. The associated surface pattern will be appended to the list of required patterns `patternList`, provided that `patternList` does not already contain this pattern.

The following code snippet demonstrates the code implementation of the DSPSA within the MSIS; the code explanation follows.

```
private void DSPSA(Vector3D rayDirection, Vector3D camPos, ArrayList patternList)
{
    double discriminant = 4 * Math.Pow(camPos * rayDirection, 2)
                        - 4 * (rayDirection * rayDirection) * (camPos * camPos)
                        + 4 * (rayDirection * rayDirection) * Math.Pow(_moon_radius, 2);

    bool encounter = false;
    double lat = 0;
    double lon = 0;

    if (discriminant < 0)
    {
        // no encounter
    }
    else if (discriminant == 0)
    {
        // one intersection
        encounter = true;
        double t = (-2 * (camPos * rayDirection)) / (2 * (rayDirection * rayDirection));

        Vector3D ray_point = camPos + rayDirection * t;
        lat = tools.rad2deg((Math.PI / 2) - Math.Acos(ray_point.z() / _moon_radius));
        lon = tools.rad2deg(Math.Atan2(ray_point.y(), ray_point.x()));
    }
```

---

[4] In case of the DSPSA, the ray origin is always the camera's location. The ray direction relates, in this context, to the position of one pixel in the later rendering on the image plane Ω.

8 Dynamical Surface Pattern Selection

```
        else
        {
            // two intersections
            encounter = true;
            double t1 = (-2 * (camPos * rayDirection) + Math.Sqrt(discriminant))
                        / (2 * (rayDirection * rayDirection));
            double t2 = (-2 * (camPos * rayDirection) - Math.Sqrt(discriminant))
                        / (2 * (rayDirection * rayDirection));

            Vector3D ray_point1 = camPos + rayDirection * t1;
            Vector3D ray_point2 = camPos + rayDirection * t2;
            double ray_norm1 = (ray_point1 - camPos).norm();
            double ray_norm2 = (ray_point2 - camPos).norm();

            if (ray_norm1 < ray_norm2)
            {
                lat = tools.rad2deg((Math.PI / 2) - Math.Acos(ray_point1.z() / _moon_radius));
                lon = tools.rad2deg(Math.Atan2(ray_point1.y(), ray_point1.x()));
            }
            else
            {
                lat = tools.rad2deg((Math.PI / 2) - Math.Acos(ray_point2.z() / _moon_radius));
                lon = tools.rad2deg(Math.Atan2(ray_point2.y(), ray_point2.x()));
            }
        }

        if (lon < 0)
        {
            lon += 360;
        }

        if (encounter)
        {
            double lat_start = 0;
            double lat_end = 0;
            double lon_start = 0;
            double lon_end = 0;
            if (lat >= 0)
            {
                lat_start = Math.Truncate(lat / 5) * 5;
                lat_end = lat_start + 5;
            }
            else
            {
                lat_end = Math.Truncate(lat / 5) * 5;
                lat_start = lat_end - 5;
            }
```

8.3 MSIS Implementation

```
            lon_start = Math.Truncate(lon / 5) * 5;
            lon_end = lon_start + 5;

            string temp = "_lat_" + lat_start + "_" + lat_end + "_lon_"
                        + lon_start + "_" + lon_end;
            if (!patternList.Contains(temp))
            {
                patternList.Add(temp);
            }
        }
    }
```

Listing 8.1 MSIS code implementation of the Dynamical Surface Pattern Selection Algorithm (DSPSA).

Based on the user-given camera position (`camPos`) and ray direction (`rayDirection`), the discriminant of the quadratic formula (cf. equation 8.4 on page 105) is calculated (lines 3–5 of listing 8.1), whereas **o** is the camera position and **d** the ray direction. Subsequently, its value is evaluated at the lines 11–49. If the discriminant is less than 0, the Moon has not been hit by the ray. The execution of the DSPSA ends here, since no surface pattern has to be added to the `patternList`.

If the discriminant evaluates to 0, the ray is tangential to one point on the Moon's surface. In this case, the value of the auxiliary variable `encounter` is set to `true` (line 18), which enables the addition of the surface pattern hit to the `patternList` at the end of the method. The value of the ray equation parameter λ (see equation 8.4), which is named here as variable `t`, is calculated (line 19). An evaluation of the ray equation with the determined parameter now yields the hit point of the ray on the Moon's surface (line 21). Using the equations out of section 3.2, the selenographic coordinates (latitude and longitude) of this point are calculated (lines 22–23).

If the discriminant is greater than 0, the ray intersects the Moon twice. `encounter` is likewise set to `true` (line 28). The evaluation of equation 8.4 nows yields two values for the ray equation parameter λ (lines 29–32). The surface hit points `ray_point1` and `ray_point2` are calculated (lines 34 and 35). To decide, which of both points is visible on the later rendered image (one point is on the camera-averted side), the 2-norm of the vectors `ray_point1 - camPos` and `ray_point2 - camPos`, which is the distance of the two points with reference to the camera's location, is computed and compared. The point with the smallest distance to the camera will be visible on the later rendered image, so the selenographic coordinates for this point will be determined (lines 36–48).

Subsequently, the determined longitude of the visible surface point will be adjusted to be in $[0, 360°)$ (lines 51–54), since `atan2` yields values in the range $[-180°, 180°]$. Using the selenographic coordinates of the surface hit point, the file name of the associated surface pattern will be ascertained (lines 58–78) and appended as string to the `patternList` (lines 79–82). The DSPSA has finished with this step.

Within the `Simulation` class, the DSPSA is invoked using the following code snippet:

```
ArrayList pattern = new ArrayList();
for (double y = 1; y <= this._height; y++)
{
    for (double x = 1; x <= this._width; x++)
    {
        Vector3D rayDirection = sc.getPOVDirection() + (((1 + this._height - (2 * y))
                              / (2 * this._height)) * sc.getPOVUp())
                              - (((1 + this._width - (2 * x)) / (2 * this._width))
                              * sc.getPOVRight());
        this.DSPSA(rayDirection, scPos, pattern);
    }
}
```

Listing 8.2 DPSPA invocation within the `Simulation` class. This code snippet has to be executed for each image to be rendered.

First, an empty variable for the list of pattern filenames, `pattern`, is instantiated (line 1). For each single pixel of the later rendered image, the ray direction for this particular pixel will be calculated using equation 8.15 on page 108; the DSPSA method is now invoked specifying this information (lines 2–12).

8.4 Drawbacks of this Method

The chosen method assumes the Moon to be a perfect sphere with a radius of $r_{\mathbb{C}} = 1.73715 \cdot 10^6 \ (\pm 10)$ m. By implication, this method does not take the Moon's local topography into account. An MSISF database query[5] for the highest and lowest elevation of the Moon's surface gives values of 1 748 160.5 m and 1 728 278 m (both values are the planetary radius at this particular surface point), respectively, which means an absolute altitude difference of 19.8825 km across the Moon's surface. Neglecting this topography means that the DSPSA may produce false-positive (pixels identified to show a part of the Moon's surface, but not showing in reality) and false-negative (pixels identified not to show a part of the Moon's surface, but showing in reality) results in the periphery of the Moon's surface. This may have influence on renderings

---

[5] `SELECT min(planetary_radius), max(planetary_radius) FROM ldem_64;`

under rendering conditions with low flight altitudes (below 100–200 km) in conjuction with a non-nadir facing view direction.

Additionally, objects which are outside the visible area of the rendered image can have an influence on renderings, for example, by (missing) shadow casting. It is possible that an object is not visible on the rendering itself, but its shadow is visible (see figure 8.4 for an example). This effect will have increased influence on renderings near the poles, since the patterns are squeezed together here and so a lot more patterns have to be selected for one rendering in comparsion with a rendering using the same conditions (flight altitude, view direction) near the equator.

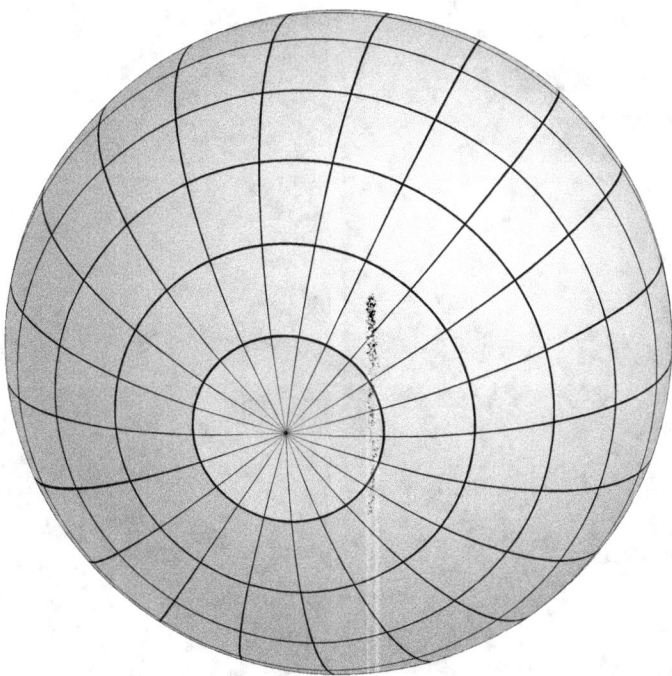

Figure 8.3 Illustration of pattern squeezing at the poles (the orange marked area): A rendering of the same surface area near the poles needs more surface patterns. The illustration is not drawn to scale (the MSISF surface patterns are much smaller).

However, accounting for the true Moon's topography within a pattern selection algorithm would be extremely demanding. The MSIS-implemented algorithm has been designed to perform and scale very well, even for large renderings (the DSPSA took up 2–5 seconds for images in resolutions of about 2.3 mega pixels in the tests).

Although these drawbacks are known, they do not represent issues for the thesis' purpose; but they could be an issue for other purposes.

8 Dynamical Surface Pattern Selection

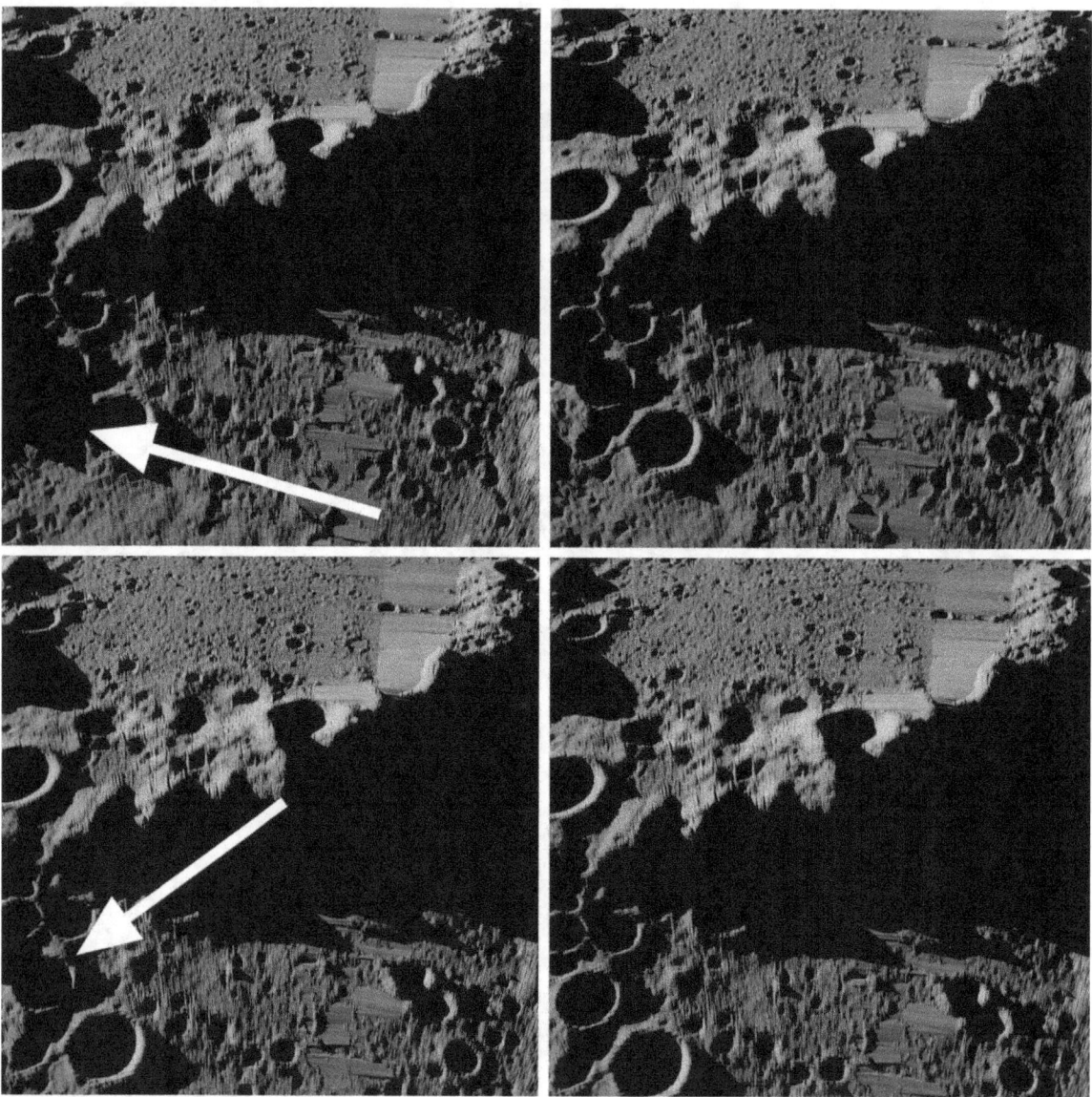

Figure 8.4 DSPSA drawbacks visible on successive renderings of a series. The time is frozen (the Sun's position in relation to the Moon's position will not change), but the camera moves along a trajectory around the Moon. The camera moves to the image north with each picture (from left to right, top to bottom).

Each left picture shows a shadow caused by an object, which is not visible in the scene. Each right picture shows the next rendering, when the camera has moved a little bit upwards — some shadows disappear, because the patterns, on which the objects causing the shadows are located, are not selected anymore. The object, never visible itself, has disappeared, and so has its shadow.

XML Meta Rendering Information and Rendering Annotations

9.1 Definition of the MSIS Output

The MSIS is designed to produce

- a rendering of the Moon's surface using user-supplied scene settings (simulation time, orbit, orientation, camera settings etc.) as PNG (*Portable Network Graphics*) image (`MoonSurfIllumSim_step_XXXXX.png`),

- the corresponding POV-Ray file, containing all instructions for generating the actual scene with POV-Ray, (`MoonSurfIllumSim_step_XXXXX.pov`) and

- an XML (*Extensible Markup Language*) file with meta information of several global rendering parameters, as well as the local solar illumination angle of particular surface points (`MoonSurfIllumSim_step_XXXXX.xml`)

as basic output in fulfillment of the primary thesis objective.

Notwithstanding that this output is sufficient in the light of the thesis' task definition, an extended output, pursuing the visualization of the XML meta information, has been envisaged. This operation mode can be activated with the command-line switch `--rendering-annotation` and produces an additional PNG file (`MoonSurfIllumSim_step_XXXXX.annotated.png`). In this PNG file, the local solar illumination angle, which is given in the pixel-wise information block in the XML file, will be visualized as a plot overlay of the resulting vector field. In addition, some general rendering information will be drawn as text in the upper left corner of the rendering. An example of an annotated rendering is given in figure 9.1.

This annotated rendering will make a quick visualization of the XML file meta information contents possible for the user. For all further applications, additional software capable of using the generated XML file can be developed. It would be conceivable, for example, to develop a rendering viewer which shows the local solar illumination angle as a text value according to the current mouse pointer position on the rendering. Such a software could easily parse and import the produced XML meta information file of a rendering using the MSIS-supplied XML DTD (*Document Type Definition*).

9.2 Structure of the XML Meta Information File

The XML file format has been chosen deliberately, as it guarantees an easy, flawless, extensible and interoperable way of storing and exchanging the auxiliary rendering meta information with other applications. The MSIS will produce a well-formed and valid XML file, which is separated into two distinct classes: A general information block provides particulars on rendering parameters with a global scope, including the

- simulation time as MJD and UTC date,
- camera position and orientation,
- Sun position,
- flight altitude,
- LDEM resolution used,
- FOV,
- MSIS version and
- command line arguments used,

while the pixel information block gives information on

- the shown local solar illumination angle and
- the surface location

of single rendering pixels. A DTD, containing a definition of all used elements, attributes and grammatical rules, is supplied along with the MSISF (`/lib/MSISRendering.dtd`), which allows proper validation and parsing of the XML file (see appendix B.2). Special attention was devoted

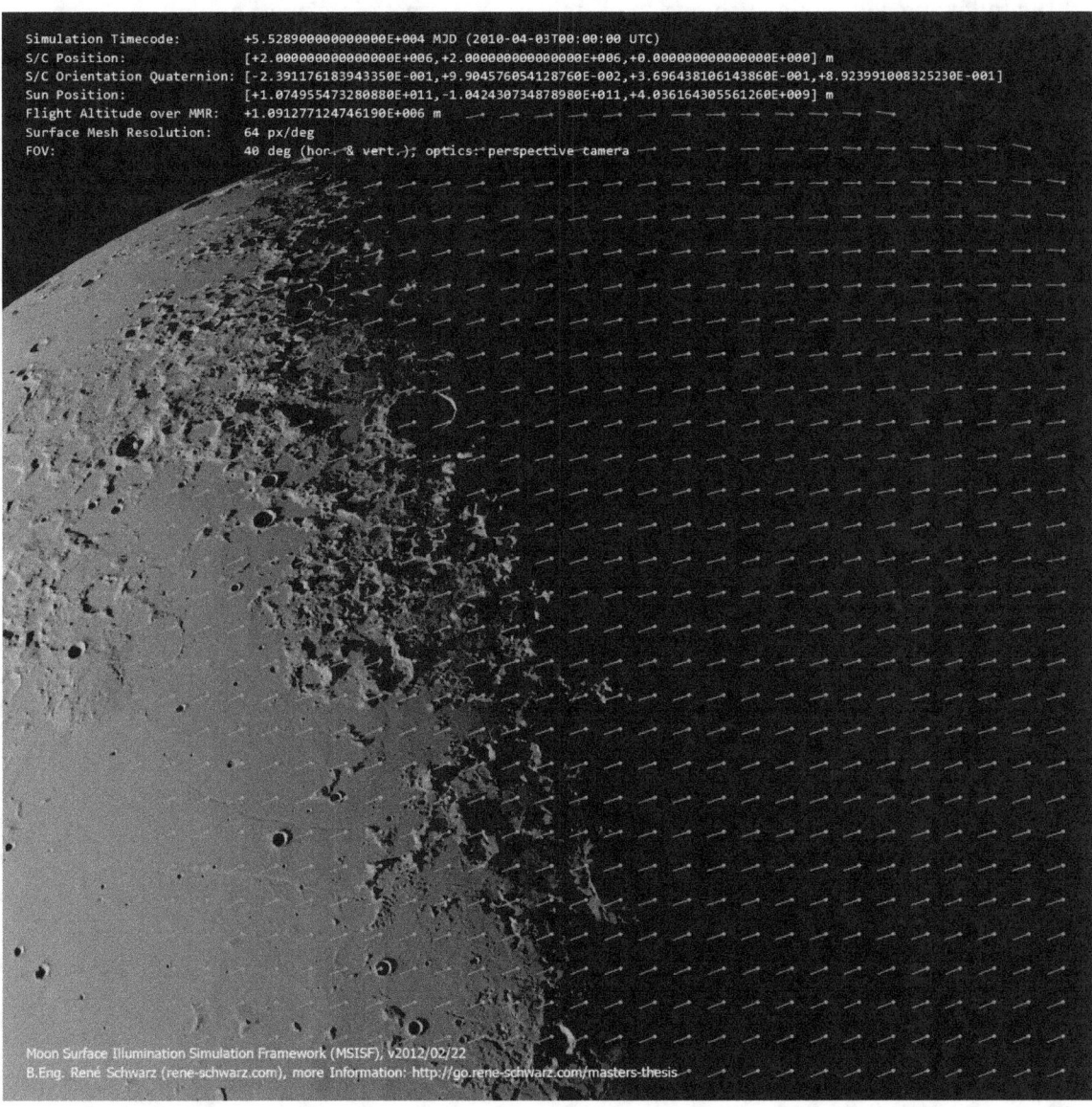

Figure 9.1 MSIS rendering with rendering annotations activated. The red lines indicate the local solar illumination angle for each grid sample point, which is marked as a red dot at the beginning of the lines. The MSIS will only visualize the illumination angle on pixels, which show the Moon's surface (not the space background).

9 XML Meta Rendering Information and Rendering Annotations

to flexible extendability of the XML file through MSIS code changes, for example, to represent the output in different units.

The following code snippet shows an exemplary XML meta information file[1]:

```xml
<?xml version="1.0" encoding="UTF-8" standalone="no" ?>
<!DOCTYPE MSISRendering SYSTEM "../lib/MSISRendering.dtd">
<MSISRendering>
    <GeneralInformation>
        <SimulationTime>
            <MJD>+5.528900000000000E+004</MJD>
            <UTC>2010-04-03T00:00:00Z</UTC>
        </SimulationTime>
        <CameraPosition unit="m">
            <Vector3D x="+2.000000000000000E+006" y="+2.000000000000000E+006"
            z="+0.000000000000000E+000" />
        </CameraPosition>
        <CameraOrientation>
            <Quaternion r="+8.923991008325230E-001" x="-2.391176183943350E-001"
            y="+9.904576054128760E-002" z="+3.696438106143860E-001" />
        </CameraOrientation>
        <SunPosition unit="m">
            <Vector3D x="+1.074955473280880E+011" y="-1.042430734878980E+011"
            z="+4.036164305561260E+009" />
        </SunPosition>
        <FlightAltitude unit="m">+1.091277124746190E+006</FlightAltitude>
        <SurfaceResolution unit="px/deg">64</SurfaceResolution>
        <FOV unit="deg">40</FOV>
        <MSISVersion>2012/02/22</MSISVersion>
        <CommandLine>--fixed-state {55289,2E6,2E6,0,-2.391176183943345e-001,▼
        +9.904576054128762e-002,+3.696438106143861e-001,+8.923991008325228e-001}▼
        --batch --rendering-annotation -r 64</CommandLine>
    </GeneralInformation>
    <PixelInformation>
        <Pixel h="450" v="100">
            <SelenographicCoordinates lat="+4.458440049692400E+001"
            lon="+5.679090635326980E+001" units="deg" />
            <IlluminationDirection unit="deg">+2.808505864577170E+002</IlluminationDirection>
        </Pixel>
        <Pixel h="500" v="100">
            <SelenographicCoordinates lat="+4.165137059729800E+001"
            lon="+5.868035440967800E+001" units="deg" />
            <IlluminationDirection unit="deg">+2.790328202771270E+002</IlluminationDirection>
        </Pixel>
```

---

[1] The symbol ▼ means that a line break was introduced for typographic reasons, but there must be no line break here, because it would cause a malfunction (command line arguments, for example, never contain line breaks).

```
40              ...
41              ...
42          </PixelInformation>
43      </MSISRendering>
```

Lines 1 and 2 represent the XML and doctype declaration, which gives meta information about the XML file itself (encoding, version, DTD used). These lines are followed by the `MSISRendering` tag, which marks the body of the XML document. All pre-defined tags and attributes will be explained within the next paragraphs. By convention, all numbers are displayed and stored in the MATLAB `longE` format (floating-point numbers in scientific notation with 15 digits after the decimal point).

CameraOrientation
Camera rotation/orientation quaternion (as specified in chapter 7).

Attributes:	none
Element Contents:	only defined child elements
Possible Child Elements:	`Quaternion`

CameraPosition
The camera/spacecraft position in the ME/PA reference frame.

Attributes:	`unit`	unit of the numerical value given in the child element [string]
Element Contents:	only defined child elements	
Possible Child Elements:	`Vector3D`	

CommandLine
All user-given command line arguments for the current MSIS call.

Attributes:	none
Element Contents:	string
Possible Child Elements:	none

FlightAltitude
Flight altitude over MMR.

Attributes:	`unit`	unit of the numerical value [string]
Element Contents:	numerical value	
Possible Child Elements:	none	

GeneralInformation
Contains information on general rendering parameters (valid for all rendering pixels).

9 XML Meta Rendering Information and Rendering Annotations

Attributes:	none
Element Contents:	only defined child elements
Possible Child Elements:	`SimulationTime`, `CameraPosition`, `CameraOrientation`, `SunPosition`, `FlightAlititude`, `SurfaceResolution`, `FOV`, `MSISVersion`, `CommandLine`

IlluminationDirection

Local solar illumination angle for a certain pixel (cf. section 9.3).

Attributes:	`unit`	unit of the numerical value [string]
Element Contents:	numerical value	
Possible Child Elements:	none	

MJD

A UTC date given as Modified Julian Date.

Attributes:	none
Element Contents:	numerical value
Possible Child Elements:	none

MSISRendering

Contains the two general information classes and serves as body for the XML document.

Attributes:	none
Element Contents:	only defined child elements
Possible Child Elements:	`GeneralInformation`, `PixelInformation`

MSISVersion

MSIS version used.

Attributes:	none
Element Contents:	MSIS version string
Possible Child Elements:	none

Pixel

Container for information on a certain rendering pixel.

Attributes:	`h`	horizontal pixel location [uint]
	`v`	vertical pixel location [uint]
Element Contents:	only defined child elements	
Possible Child Elements:	`SelenographicCoordinates`, `IlluminationDirection`	

PixelInformation

Contains local information on certain pixels of the rendering.

Attributes:	none
Element Contents:	only defined child elements
Possible Child Elements:	`Pixel`

Quaternion

A quaternion representing a spatial rotation/orientation in the ME/PA reference frame (cf. chapter 7).

Attributes:	`r`	real part of the quaternion [double]
	`x`	vector part x-component [double]
	`y`	vector part y-component [double]
	`z`	vector part z-component [double]
Element Contents:	none	
Possible Child Elements:	none	

SelenographicCoordinates

A selenographic coordinate (latitude/longitude).

Attributes:	`lat`	selenographic latitude [double]
	`lon`	selenographic longitutde [double]
	`units`	units of the latitude and longitude [string]
Element Contents:	none	
Possible Child Elements:	none	

SimulationTime

The rendering simulation time. This element is just a container for other elements.

Attributes:	none
Element Contents:	only defined child elements
Possible Child Elements:	`MJD, UTC`

SunPosition

The Sun position in the ME/PA reference frame.

Attributes:	`unit`	unit of the numerical value given in the child element [string]
Element Contents:	only defined child elements	
Possible Child Elements:	`Vector3D`	

UTC

A UTC date given as ISO 8601 string.

Attributes:	none

Element Contents:	full ISO 8601 string
Possible Child Elements:	none

Vector3D

A vector representing the position of a point in the ME/PA reference frame.

Attributes:	x	vector x-component [double]
	y	vector y-component [double]
	z	vector z-component [double]
Element Contents:	none	
Possible Child Elements:	none	

9.3 Determination of the Local Solar Illumination Angle

The determination and output of the local solar illumination angle is a non-trival task, since this angle is varying for each pixel of a rendering. Because of these variations, the local solar illumination angle has to be determined and listed in the meta information file, usually for each single pixel, which would result in a huge and unmanageable XML file. Fortunately, higher tolerances are possible, as the variations will likely be in a low order of magnitude between adjacent pixels unless the flight altitude is really high.

For these reasons, the MSIS will generate pixel meta information by default only in vertical and horizontal margins of 50 px, which results in a gridded discretization. The horizontal and vertical spacing between the sample points can be user-defined[2]. For each sample point it will be proved, whether this actual pixel displays either the Moon's surface or the space background. If a pixel represents the Moon's surface, the following information will be generated:

- Pixel position within the rendering

- Position of the shown surface point on the Moon in selenographic coordinates (latitude/longitude)

- Local solar illumination angle

The procedure for determining the local illumination angle for each grid point (s_{x1}, s_{y1}) on the image plane Ω is as follows (cf. figure 9.2):

---

[2] Complete meta information for each single pixel will be generated by setting the horizontal and vertical spacing to 1 px. This is not recommended.

9.3 Determination of the Local Solar Illumination Angle

1. Shoot one ray $\mathcal{L}_1$ starting from the camera's focal point $\mathbf{c}_{\text{pos}}$ through the grid point (s_{x1}, s_{y1}) on the image plane Ω: Does the ray hit the Moon's surface? If yes, continue. Otherwise choose next grid point.

2. Obtain the surface hit point $\mathbf{p}_{\text{surf}}$ in rectangular and selenographic coordinates.

3. Calculate the Sun direction $\hat{\mathbf{d}}_\odot$ (unit vector between surface hit point and Sun position).

4. Calculate the local tangent plane Ξ of the surface hit point.

5. Obtain the subsurface point of $\mathbf{p}_{\text{surf}} + 1\,000\,\hat{\mathbf{d}}_\odot$ on the tangent plane Ξ using a projection technique. This point is named local illumination point $\mathbf{p}_{\text{local}}$.

6. Calculate the local illumination direction $\hat{\mathbf{d}}_{\text{local}}$ (unit vector between focal point and local illumination point).

7. Calculate the intersection between the line $\mathcal{L}_2: \mathbb{R} \to \mathbb{R}^3, \lambda \mapsto \mathbf{c}_{\text{pos}} + \lambda\,\hat{\mathbf{d}}_{\text{local}}$ and image plane in rendering coordinates (s_{x2}, s_{y2}).

8. Compute the angle between the vectors $(0, -1000)^\text{T}$ and $(s_{x2} - s_{x1}, s_{y2} - s_{y1})^\text{T}$. This angle is the local solar illumination angle α.

The next sections discuss all theoretical steps in detail. Additionally, code snippets show the MSIS implementation of the particular step. The algorithm is implemented by the function `getIlluminationDirectionPerRenderingPixel(SpacecraftState sc, uint x, uint y)` into the MSIS, which expects the pixel to be considered with its coordinates x and y as arguments, as well as an object of the `SpacecraftState` class, which contains information about the scene geometry.

Step 1: Visibility Test

First, it has to be proved whether the given rendering pixel shows the Moon's surface or the space background. This is determined using the ray tracing algorithm of the DSPSA (c.f. chapter 8): A ray

$$\mathcal{L}_1: \mathbb{R} \to \mathbb{R}^3, \lambda \mapsto \underbrace{\mathbf{c}_{\text{pos}}}_{=\mathbf{o}} + \lambda\,\underbrace{\hat{\mathbf{d}}(s_{x1}, s_{y1})}_{=\mathbf{d}} \qquad (9.1)$$

will be shot from the camera's focal point $\mathbf{c}_{\text{pos}}$ in the direction $\hat{\mathbf{d}}(s_{x1}, s_{y1})$. This direction takes the internal camera geometry into account to ensure that the ray will go through the given pixel (s_{x1}, s_{y1}) on the image plane Ω (see equation 8.15). Using the ray tracing technique in section 8.1, it can be calculated if the ray will hit the Moon's surface or not by evaluating equation 8.4.

9 XML Meta Rendering Information and Rendering Annotations

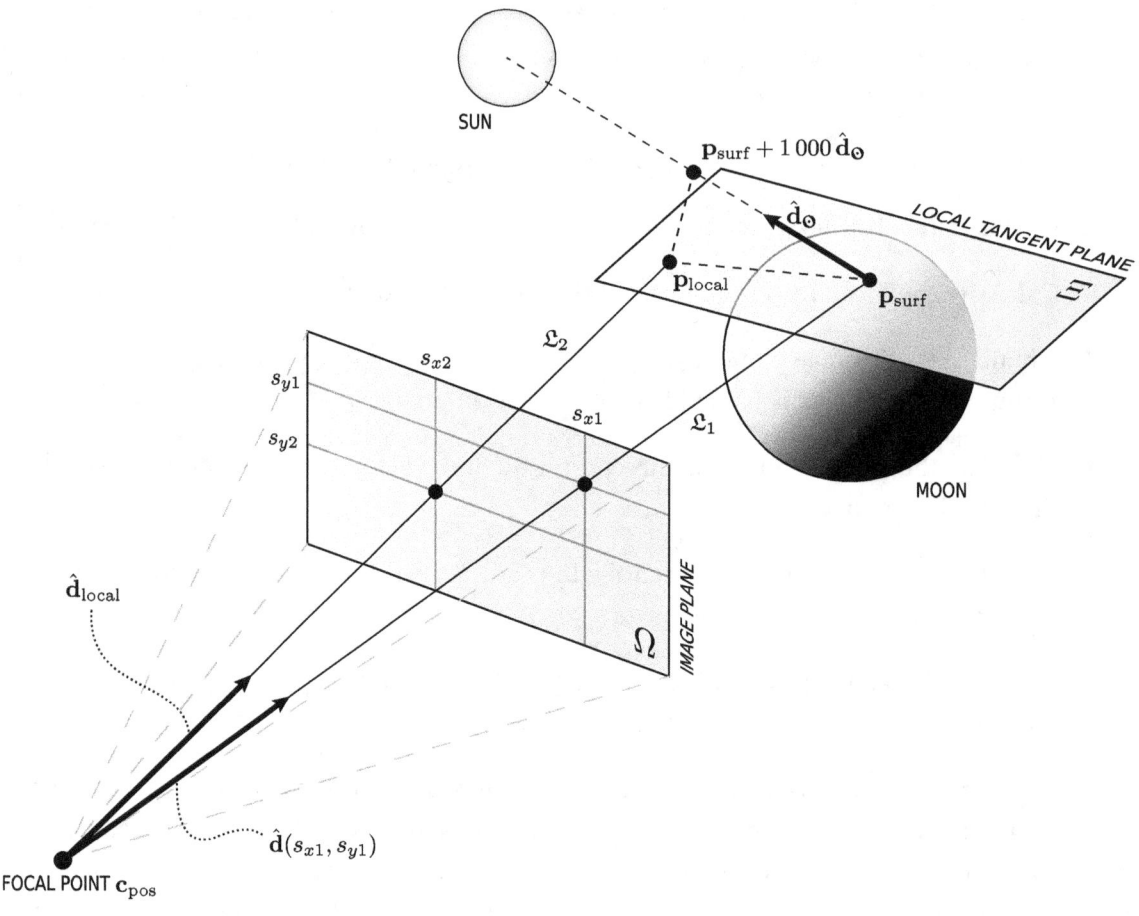

Figure 9.2 Schematic of the local illumination angle determination principle.

If the Moon's surface is hit, the algorithm continues with step 2. Otherwise, the execution of the algorithm for this certain pixel ends and the next grid sample point is chosen.

```
────────────────────── MSIS Implementation ──────────────────────
// STEP 1: Visibility Test
Vector3D rayDirection = (sc.getPOVDirection() + (((1 + Convert.ToDouble(this._height)
                       - (2 * Convert.ToDouble(y))) / (2 * Convert.ToDouble(this._height)))
                       * sc.getPOVUp()) - (((1 + Convert.ToDouble(this._width)
                       - (2 * Convert.ToDouble(x))) / (2 * Convert.ToDouble(this._width)))
                       * sc.getPOVRight())).unit();
double discriminant = 4 * Math.Pow(c_pos * rayDirection, 2)
                    - 4 * (rayDirection * rayDirection) * (c_pos * c_pos)
                    + 4 * (rayDirection * rayDirection) * Math.Pow(_moon_radius, 2);

bool encounter = false;
```

9.3 Determination of the Local Solar Illumination Angle

```
12  Vector3D p_surf = new Vector3D();
13
14  if (discriminant < 0)
15  {
16      // no encounter
17  }
18  else if (discriminant == 0)
19  {
20      // one intersection
21      encounter = true;
22      double t = (-2 * (c_pos * rayDirection)) / (2 * (rayDirection * rayDirection));
23
24      p_surf = c_pos + rayDirection * t;
25  }
26  else
27  {
28      // two intersections
29      encounter = true;
30      double t1 = (-2 * (c_pos * rayDirection) + Math.Sqrt(discriminant))
31              / (2 * (rayDirection * rayDirection));
32      double t2 = (-2 * (c_pos * rayDirection) - Math.Sqrt(discriminant))
33              / (2 * (rayDirection * rayDirection));
34
35      Vector3D ray_point1 = c_pos + rayDirection * t1;
36      Vector3D ray_point2 = c_pos + rayDirection * t2;
37      double ray_norm1 = (ray_point1 - c_pos).norm();
38      double ray_norm2 = (ray_point2 - c_pos).norm();
39
40      if (ray_norm1 < ray_norm2)
41      {
42          p_surf = ray_point1;
43      }
44      else
45      {
46          p_surf = ray_point2;
47      }
48  }
49  if (encounter)
50  {
51      // Pixel shows the Moon's surface: Continue with executing steps 2 - 8.
52  }
53  else
54  {
55      // next grid sample point
56      return PixelOut;
57  }
```

9 XML Meta Rendering Information and Rendering Annotations

Step 2: Obtaining the Surface Hit Point

The position of the shown surface hit point can easily be obtained from the ray tracing information, as explained in section 8.1: By evaluating the ray equation with parameter λ, the spatial position of the ray hit point can be determined. These rectangular coordinates can be converted into selenographical coordinates using their definition (see section 3.2)

$$\vartheta = \frac{\pi}{2} - \arccos \frac{p_{\text{surf},z}}{r_{\text{☾}}} \quad \text{and} \tag{9.2}$$

$$\varphi = \arctan2(p_{\text{surf},y}, p_{\text{surf},x}), \tag{9.3}$$

where ϑ is the latitude and φ the longitude; arctan2 is the two-argument arctangent function[3]. The rendering coordinates (s_{x1}, s_{y1}) are given implicitly by the grid definition and the current algorithm iteration.

─────────── MSIS Implementation ───────────

```
double x1 = Convert.ToDouble(x);
double y1 = Convert.ToDouble(y);

// STEP 2: Obtaining the Surface Hit Point
PixelOut.lat = tools.rad2deg((Math.PI / 2) - Math.Acos(p_surf.z() / _moon_radius));
PixelOut.lon = tools.rad2deg(Math.Atan2(p_surf.y(), p_surf.x()));
```

Step 3: Calculation of the Sun's Direction

For each surface point $\mathbf{p}_{\text{surface}}$ identified with the aforementioned ray tracing technique, the normalized direction[4] $\hat{\mathbf{d}}_{\odot}$ of the Sun's position as seen from that point will be determined by subtracting the surface point $\mathbf{p}_{\text{surface}}$ from the Sun's position $\mathbf{p}_{\odot}$ and a subsequent normalization:

$$\hat{\mathbf{d}}_{\odot} = \frac{\mathbf{p}_{\odot} - \mathbf{p}_{\text{surf}}}{\|\mathbf{p}_{\odot} - \mathbf{p}_{\text{surf}}\|} \tag{9.4}$$

---

[3] $\arctan2(y, x) = \begin{cases} \arctan\left(\frac{y}{x}\right) & x > 0 \\ \arctan\left(\frac{y}{x}\right) + \pi & y \geq 0, x < 0 \\ \arctan\left(\frac{y}{x}\right) - \pi & y < 0, x < 0 \\ +\frac{\pi}{2} & y > 0, x = 0 \\ -\frac{\pi}{2} & y < 0, x = 0 \\ \text{undefined} & y = 0, x = 0 \end{cases}$

[4] Normalized, in this context, means the construction of a unit vector.

```
─────────────────────────── MSIS Implementation ───────────────────────────
1  // STEP 3: Calculation of the Sun's direction
2  Vector3D hat_p_surf = p_surf.unit();
3  Vector3D hat_d_sun = (sc.getSunPosition() - p_surf).unit();
```

Step 4: Derivation of the Local Tangent Plane of the Surface Hit Point

The local tangent plane Ξ of the surface hit point $\mathbf{p}_{\text{surf}}$ is defined as the plane, which contains all tangents of the surface hit point $\mathbf{p}_{\text{surf}}$ on the Moon's sphere. Ξ is necessarily perpendicular to the position vector of the surface hit point $\mathbf{p}_{\text{surf}}$, which originates in the Moon's center. This way, the unit vector $\hat{\mathbf{p}}_{\text{surf}}$ of $\mathbf{p}_{\text{surf}}$ is the normal vector $\hat{\mathbf{n}}_\Xi$ of Ξ. In order to construct a plane, two non-collinear support vectors $\mathbf{u}_1$ and $\mathbf{u}_2$ need to be defined. A support vector could be every vector which is perpendicular to the normal vector $\hat{\mathbf{n}}_\Xi$ of Ξ, satisfying

$$\langle \hat{\mathbf{n}}_\Xi, \mathbf{u}_1 \rangle = 0 \text{ and } \langle \hat{\mathbf{n}}_\Xi, \mathbf{u}_2 \rangle = 0. \tag{9.5}$$

This is obviously true[5] for both arbitrarily chosen support vectors

$$\mathbf{u}_1 = (-\hat{p}_{\text{surf},y}, \hat{p}_{\text{surf},x}, 0)^{\text{T}} \text{ and } \mathbf{u}_2 = (0, -\hat{p}_{\text{surf},z}, \hat{p}_{\text{surf},y})^{\text{T}}. \tag{9.6}$$

The local tangent plane Ξ can now be written as

$$\Xi(\lambda_1, \lambda_2) \colon \mathbb{R} \times \mathbb{R} \to \mathbb{R}^3, (\lambda_1, \lambda_2) \mapsto \mathbf{p}_{\text{surf}} + \lambda_1 \begin{pmatrix} -\hat{p}_{\text{surf},y} \\ \hat{p}_{\text{surf},x} \\ 0 \end{pmatrix} + \lambda_2 \begin{pmatrix} 0 \\ -\hat{p}_{\text{surf},z} \\ \hat{p}_{\text{surf},y} \end{pmatrix}. \tag{9.7}$$

Step 5: Determination of a Subsurface Point of the Solar Illumination Direction on the Local Tangent Plane

The Sun's illumination direction line

$$\mathcal{L}_\odot(\lambda_3) \colon \mathbb{R} \to \mathbb{R}^3, \lambda_3 \mapsto \mathbf{p}_{\text{surf}} + \lambda_3 \hat{\mathbf{d}}_\odot \tag{9.8}$$

has to be projected to the local tangent plane Ξ of $\mathbf{p}_{\text{surf}}$ to obtain the local solar illumination line

$$\mathcal{L}_{\text{local}}(\lambda_4) \subset \Xi \colon \mathbb{R} \to \mathbb{R}^3, \lambda_4 \mapsto \mathbf{p}_{\text{surf}} + \lambda_4 (\mathbf{p}_{\text{local}} - \mathbf{p}_{\text{surf}}) \tag{9.9}$$

along the Moon's surface. Due to the implicit knowledge of the intersection point of the tangential plane Ξ and the line defined by the Sun's position and the surface hit point — namely, the surface hit point $\mathbf{p}_{\text{surf}}$ itself — it is sufficient to determine a single point $\mathbf{p}_{\text{local}}$ of the local solar illumination line. Figure 9.3 gives a detailed overview of the geometrical constitution.

---

[5] $\langle \hat{\mathbf{n}}_\Xi, \mathbf{u}_1 \rangle = \langle (\hat{p}_{\text{surf},x}, \hat{p}_{\text{surf},y}, \hat{p}_{\text{surf},z})^{\text{T}}, (-\hat{p}_{\text{surf},y}, \hat{p}_{\text{surf},x}, 0)^{\text{T}} \rangle = \hat{p}_{\text{surf},x} \hat{p}_{\text{surf},y} - \hat{p}_{\text{surf},x} \hat{p}_{\text{surf},y} + 0 = 0$
$\langle \hat{\mathbf{n}}_\Xi, \mathbf{u}_2 \rangle = \langle (\hat{p}_{\text{surf},x}, \hat{p}_{\text{surf},y}, \hat{p}_{\text{surf},z})^{\text{T}}, (0, -\hat{p}_{\text{surf},z}, \hat{p}_{\text{surf},y})^{\text{T}} \rangle = 0 - \hat{p}_{\text{surf},y} \hat{p}_{\text{surf},z} + \hat{p}_{\text{surf},y} \hat{p}_{\text{surf},z} = 0$

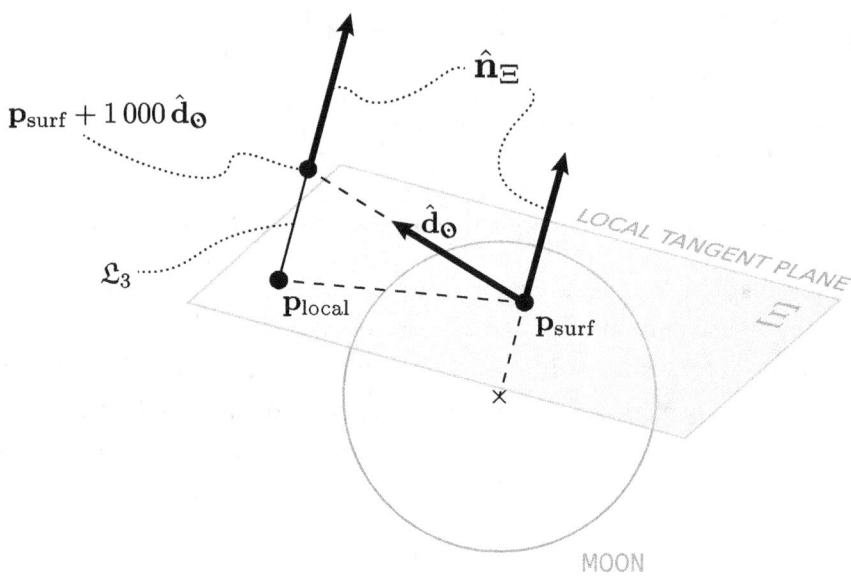

Figure 9.3 Geometrical construction of $\mathbf{p}_{\text{local}}$.

The aforementioned point can be the projection of one arbitrarily chosen point of $\mathfrak{L}_\oplus$ to Ξ; it is called $\mathbf{p}_{\text{local}}$. To ensure an adequate margin between $\mathbf{p}_{\text{surf}}$ and $\mathbf{p}_{\text{local}}$, a margin of $\lambda_3 = 1\,000$ is found to be suitable. As elucidated in step 4, the normal vector $\hat{\mathbf{n}}_\Xi$ of the local tangent plane Ξ is the unit vector $\hat{\mathbf{p}}_{\text{surf}}$ of the surface hit point $\mathbf{p}_{\text{surf}}$. This way, $\hat{\mathbf{n}}_\Xi$ must be the direction vector of a line

$$\mathfrak{L}_3(\lambda_5) \colon \mathbb{R} \to \mathbb{R}^3, \lambda_5 \mapsto \mathfrak{L}_\oplus(\lambda_3 = 1\,000) + \lambda_5\,\hat{\mathbf{n}}_\Xi = \mathbf{p}_{\text{surf}} + 1\,000\,\hat{\mathbf{d}}_\oplus + \lambda_5\,\hat{\mathbf{n}}_\Xi \qquad (9.10)$$

through $\mathfrak{L}_\oplus(\lambda_3 = 1\,000)$, which is perpendicular to the tangential plane Ξ, too. The subsurface point $\mathbf{p}_{\text{surf}}$ is the intersection point of $\mathfrak{L}_3$ and Ξ and can be obtained by equating $\mathfrak{L}_3$ and Ξ

$$\mathfrak{L}_3(\lambda_5) = \Xi(\lambda_1, \lambda_2)$$
$$\mathbf{p}_{\text{surf}} + 1\,000\,\hat{\mathbf{d}}_\oplus + \lambda_5\,\hat{\mathbf{n}}_\Xi = \mathbf{p}_{\text{surf}} + \lambda_1\,\mathbf{u}_1 + \lambda_2\,\mathbf{u}_2$$
$$1\,000 \begin{pmatrix} \hat{d}_{\oplus,x} \\ \hat{d}_{\oplus,y} \\ \hat{d}_{\oplus,z} \end{pmatrix} + \lambda_5 \begin{pmatrix} \hat{p}_{\text{surf},x} \\ \hat{p}_{\text{surf},y} \\ \hat{p}_{\text{surf},z} \end{pmatrix} = \lambda_1 \begin{pmatrix} -\hat{p}_{\text{surf},y} \\ \hat{p}_{\text{surf},x} \\ 0 \end{pmatrix} + \lambda_2 \begin{pmatrix} 0 \\ -\hat{p}_{\text{surf},z} \\ \hat{p}_{\text{surf},y} \end{pmatrix} \qquad (9.11)$$

and solving for λ_5

$$\lambda_5 = -\frac{1\,000\left(\hat{d}_{\oplus,x}\hat{p}_{\text{surf},x} + \hat{d}_{\oplus,y}\hat{p}_{\text{surf},y} + \hat{d}_{\oplus,z}\hat{p}_{\text{surf},z}\right)}{\hat{p}^2_{\text{surf},x} + \hat{p}^2_{\text{surf},y} + \hat{p}^2_{\text{surf},z}} \qquad (9.12)$$

or λ_1 and λ_2

$$\lambda_1 = -\frac{1000\left(-\hat{d}_{\odot,y}\hat{p}_{\text{surf},x}\hat{p}_{\text{surf},y} - \hat{d}_{\odot,z}\hat{p}_{\text{surf},x}\hat{p}_{\text{surf},z} + \hat{d}_{\odot,x}\left(\hat{p}_{\text{surf},y}^2 + \hat{p}_{\text{surf},z}^2\right)\right)}{\hat{p}_{\text{surf},y}\left(\hat{p}_{\text{surf},x}^2 + \hat{p}_{\text{surf},y}^2 + \hat{p}_{\text{surf},z}^2\right)} \quad (9.13)$$

$$\lambda_2 = \frac{1000\left(\hat{d}_{\odot,z}\left(\hat{p}_{\text{surf},x}^2 + \hat{p}_{\text{surf},y}^2\right) - \left(\hat{d}_{\odot,x}\hat{p}_{\text{surf},x} + \hat{d}_{\odot,y}\hat{p}_{\text{surf},y}\right)\hat{p}_{\text{surf},z}\right)}{\hat{p}_{\text{surf},y}\left(\hat{p}_{\text{surf},x}^2 + \hat{p}_{\text{surf},y}^2 + \hat{p}_{\text{surf},z}^2\right)} \quad (9.14)$$

and a subsequent evaluation of $\mathfrak{L}_3(\lambda_5)$ or $\Xi(\lambda_1, \lambda_2)$ with these calculated values of λ_5 or λ_1 and λ_2, respectively.

```
// STEP 4: Derivation of the Local Tangent Plane of the Surface Hit Point
// (nothing to do here)
// STEP 5: Determination of a Subsurface Point of the Solar Illumination
// Direction on the Local Tangent Plane
double lambda_5 = -(1000 * (hat_d_sun.x() * hat_p_surf.x() + hat_d_sun.y() * hat_p_surf.y()
                  + hat_d_sun.z() * hat_p_surf.z())
                  ) / (
                  hat_p_surf.x() * hat_p_surf.x()
                  + hat_p_surf.y() * hat_p_surf.y()
                  + hat_p_surf.z() * hat_p_surf.z()
                  );
Vector3D p_local = p_surf + 1000 * hat_d_sun + lambda_5 * hat_p_surf;
```

Step 6: Local Illumination Direction

The local illumination direction vector $\hat{\mathbf{d}}_{\text{local}}$, which gives the direction of $\mathbf{p}_{\text{local}}$ from the previous step as seen from the camera's focal point $\mathbf{c}_{\text{pos}}$, is now built by

$$\hat{\mathbf{d}}_{\text{local}} = \frac{\mathbf{p}_{\text{local}} - \mathbf{c}_{\text{pos}}}{\|\mathbf{p}_{\text{local}} - \mathbf{c}_{\text{pos}}\|}. \quad (9.15)$$

```
// STEP 6: Local Illumination Direction
Vector3D hat_d_local = (p_local - c_pos).unit();
```

Step 7: Projection of the Local Illumination Point to the Image Plane

In this step, the projection point (s_{x2}, s_{y2}) in rendering coordinates of $\mathbf{p}_{\text{local}}$ to the camera's image plane Ω is identified. For that reason, a line

$$\mathfrak{L}_2 \colon \mathbb{R} \to \mathbb{R}^3, t \mapsto \mathbf{c}_{\text{pos}} + t\,\hat{\mathbf{d}}_{\text{local}} \tag{9.16}$$

between the camera's focal point $\mathbf{c}_{\text{pos}}$ and $\mathbf{p}_{\text{local}}$ is constructed. The intersection point of this line with the image plane Ω is calculated.

The intercept in rendering coordinates can be calculated equating the image plane

$$\Omega \colon \mathbb{R} \times \mathbb{R} \to \mathbb{R}^3, (s_x, s_y) \mapsto \mathbf{c}_{\text{pos}} + \mathbf{c}_{\text{direction}} + \frac{1 + h - 2s_y}{2h} \mathbf{c}_{\text{up}} - \frac{1 + w - 2s_x}{2w} \mathbf{c}_{\text{right}} \tag{9.17}$$

(cf. formula 8.13; description of $\mathbf{c}_{\text{direction}}$, $\mathbf{c}_{\text{up}}$, w and h op. cit.) and line $\mathfrak{L}_2$. By solving the resulting system of equations

$$\mathbf{k} \stackrel{\text{def.}}{=\!=} \mathbf{c}_{\text{pos}} + \mathbf{c}_{\text{direction}} \tag{9.18}$$

$$\mathbf{c}_{\text{pos}} + t\,\hat{\mathbf{d}}_{\text{local}} = \mathbf{k} + \frac{1 + h - 2s_{y2}}{2h} \mathbf{c}_{\text{up}} - \frac{1 + w - 2s_{x2}}{2w} \mathbf{c}_{\text{right}} \tag{9.19}$$

for the rendering coordinates s_{x2} and s_{y2}, both coordinates evaluate to

$$s_{x2} = \frac{c_1}{2c_2} \quad \text{and} \quad s_{y2} = \frac{c_3}{2c_4} \tag{9.20}$$

whereas

$$\begin{aligned}
c_1 = &- 2w c_{\text{pos},y} c_{\text{up},z} \hat{d}_{\text{local},x} - c_{\text{right},y} c_{\text{up},z} \hat{d}_{\text{local},x} - w c_{\text{right},y} c_{\text{up},z} \hat{d}_{\text{local},x} \\
&+ 2w c_{\text{pos},x} c_{\text{up},z} \hat{d}_{\text{local},y} + c_{\text{right},x} c_{\text{up},z} \hat{d}_{\text{local},y} + w c_{\text{right},x} c_{\text{up},z} \hat{d}_{\text{local},y} \\
&+ 2w c_{\text{pos},z} (c_{\text{up},y} \hat{d}_{\text{local},x} - c_{\text{up},x} \hat{d}_{\text{local},y}) \\
&+ (1 + w) c_{\text{right},z} (c_{\text{up},y} \hat{d}_{\text{local},x} - c_{\text{up},x} \hat{d}_{\text{local},y}) \\
&+ 2w c_{\text{pos},y} c_{\text{up},x} \hat{d}_{\text{local},z} + c_{\text{right},y} c_{\text{up},x} \hat{d}_{\text{local},z} + w c_{\text{right},y} c_{\text{up},x} \hat{d}_{\text{local},z} \\
&- 2w c_{\text{pos},x} c_{\text{up},y} \hat{d}_{\text{local},z} - c_{\text{right},x} c_{\text{up},y} \hat{d}_{\text{local},z} - w c_{\text{right},x} c_{\text{up},y} \hat{d}_{\text{local},z} \\
&- 2w c_{\text{up},z} \hat{d}_{\text{local},y} k_x + 2w c_{\text{up},y} \hat{d}_{\text{local},z} k_x + 2w c_{\text{up},z} \hat{d}_{\text{local},x} k_y - 2w c_{\text{up},x} \hat{d}_{\text{local},z} k_y \\
&- 2w c_{\text{up},y} \hat{d}_{\text{local},x} k_z + 2w c_{\text{up},x} \hat{d}_{\text{local},y} k_z \\
c_2 = & \, c_{\text{right},z} (c_{\text{up},y} \hat{d}_{\text{local},x} - c_{\text{up},x} \hat{d}_{\text{local},y}) \\
&+ c_{\text{right},y} (-c_{\text{up},z} \hat{d}_{\text{local},x} + c_{\text{up},x} \hat{d}_{\text{local},z}) \\
&+ c_{\text{right},x} (c_{\text{up},z} \hat{d}_{\text{local},y} - c_{\text{up},y} \hat{d}_{\text{local},z}) \\
c_3 = &- c_{\text{right},z} c_{\text{up},y} \hat{d}_{\text{local},x} - h c_{\text{right},z} c_{\text{up},y} \hat{d}_{\text{local},x} + c_{\text{right},y} c_{\text{up},z} \hat{d}_{\text{local},x}
\end{aligned}$$

9.3 Determination of the Local Solar Illumination Angle

$$
\begin{aligned}
&+ hc_{\text{right},y}c_{\text{up},z}\hat{d}_{\text{local},x} - 2hc_{\text{pos},x}c_{\text{right},z}\hat{d}_{\text{local},y} + c_{\text{right},z}c_{\text{up},x}\hat{d}_{\text{local},y} \\
&+ hc_{\text{right},z}c_{\text{up},x}\hat{d}_{\text{local},y} - c_{\text{right},x}c_{\text{up},z}\hat{d}_{\text{local},y} - hc_{\text{right},x}c_{\text{up},z}\hat{d}_{\text{local},y} \\
&+ c_{\text{pos},z}(-2hc_{\text{right},y}\hat{d}_{\text{local},x} + 2hc_{\text{right},x}\hat{d}_{\text{local},y}) + 2hc_{\text{pos},x}c_{\text{right},y}\hat{d}_{\text{local},z} \\
&- c_{\text{right},y}c_{\text{up},x}\hat{d}_{\text{local},z} - hc_{\text{right},y}c_{\text{up},x}\hat{d}_{\text{local},z} + c_{\text{right},x}c_{\text{up},y}\hat{d}_{\text{local},z} \\
&+ hc_{\text{right},x}c_{\text{up},y}\hat{d}_{\text{local},z} + 2hc_{\text{pos},y}(c_{\text{right},z}\hat{d}_{\text{local},x} - c_{\text{right},x}\hat{d}_{\text{local},z}) \\
&+ 2hc_{\text{right},z}\hat{d}_{\text{local},y}k_x - 2hc_{\text{right},y}\hat{d}_{\text{local},z}k_x - 2hc_{\text{right},z}\hat{d}_{\text{local},x}k_y \\
&+ 2hc_{\text{right},x}\hat{d}_{\text{local},z}k_y + 2hc_{\text{right},y}\hat{d}_{\text{local},x}k_z - 2hc_{\text{right},x}\hat{d}_{\text{local},y}k_z \\
c_4 =\; & c_{\text{right},z}(-c_{\text{up},y}\hat{d}_{\text{local},x} + c_{\text{up},x}\hat{d}_{\text{local},y}) \\
&+ c_{\text{right},y}(c_{\text{up},z}\hat{d}_{\text{local},x} - c_{\text{up},x}\hat{d}_{\text{local},z}) + c_{\text{right},x}(-c_{\text{up},z}\hat{d}_{\text{local},y} + c_{\text{up},y}\hat{d}_{\text{local},z}).
\end{aligned}
$$

--- MSIS Implementation ---

```
// STEP 7: Projection of the Local Illumination Point to the Image Plane
Vector3D k = c_pos + sc.getPOVDirection();
Vector3D c_up = sc.getPOVUp();
Vector3D c_right = sc.getPOVRight();
double w = Convert.ToDouble(this._width);
double h = Convert.ToDouble(this._height);

double x2 =
    (
        -2 * w * c_pos.y() * c_up.z() * hat_d_local.x() - c_right.y() * c_up.z()
            * hat_d_local.x() - w * c_right.y() * c_up.z() * hat_d_local.x()
        + 2 * w * c_pos.x() * c_up.z() * hat_d_local.y() + c_right.x() * c_up.z()
            * hat_d_local.y() + w * c_right.x() * c_up.z() * hat_d_local.y()
        + 2 * w * c_pos.z() * (c_up.y() * hat_d_local.x() - c_up.x() * hat_d_local.y())
        + (1 + w) * c_right.z() * (c_up.y() * hat_d_local.x() - c_up.x() * hat_d_local.y())
        + 2 * w * c_pos.y() * c_up.x() * hat_d_local.z() + c_right.y() * c_up.x()
            * hat_d_local.z() + w * c_right.y() * c_up.x() * hat_d_local.z()
        - 2 * w * c_pos.x() * c_up.y() * hat_d_local.z() - c_right.x() * c_up.y()
            * hat_d_local.z() - w * c_right.x() * c_up.y() * hat_d_local.z()
        - 2 * w * c_up.z() * hat_d_local.y() * k.x() + 2 * w * c_up.y() * hat_d_local.z()
            * k.x() + 2 * w * c_up.z() * hat_d_local.x() * k.y() - 2 * w * c_up.x()
            * hat_d_local.z() * k.y()
        - 2 * w * c_up.y() * hat_d_local.x() * k.z() + 2 * w * c_up.x()
            * hat_d_local.y() * k.z()
    )/(
        2 * (c_right.z() * (c_up.y() * hat_d_local.x() - c_up.x() * hat_d_local.y())
        + c_right.y() * (-c_up.z() * hat_d_local.x() + c_up.x() * hat_d_local.z())
        + c_right.x() * (c_up.z() * hat_d_local.y() - c_up.y() * hat_d_local.z()))
    );

double y2 =
```

```
    (
        -c_right.z() * c_up.y() * hat_d_local.x() - h * c_right.z() * c_up.y()
        * hat_d_local.x() + c_right.y() * c_up.z() * hat_d_local.x() + h * c_right.y()
        * c_up.z() * hat_d_local.x() - 2 * h * c_pos.x() * c_right.z() * hat_d_local.y()
        + c_right.z() * c_up.x() * hat_d_local.y() + h * c_right.z() * c_up.x()
        * hat_d_local.y() - c_right.x() * c_up.z() * hat_d_local.y() - h * c_right.x()
        * c_up.z() * hat_d_local.y() + c_pos.z() * (-2 * h * c_right.y()
        * hat_d_local.x() + 2 * h * c_right.x() * hat_d_local.y()) + 2 * h * c_pos.x()
        * c_right.y() * hat_d_local.z() - c_right.y() * c_up.x() * hat_d_local.z()
        - h * c_right.y() * c_up.x() * hat_d_local.z() + c_right.x() * c_up.y()
        * hat_d_local.z() + h * c_right.x() * c_up.y() * hat_d_local.z() + 2 * h
        * c_pos.y() * (c_right.z() * hat_d_local.x() - c_right.x() * hat_d_local.z())
        + 2 * h * c_right.z() * hat_d_local.y() * k.x() - 2 * h * c_right.y()
        * hat_d_local.z() * k.x() - 2 * h * c_right.z() * hat_d_local.x() * k.y()
        + 2 * h * c_right.x() * hat_d_local.z() * k.y() + 2 * h * c_right.y()
        * hat_d_local.x() * k.z() - 2 * h * c_right.x() * hat_d_local.y() * k.z()
    )/(
        2 * (c_right.z() * (-c_up.y() * hat_d_local.x() + c_up.x() * hat_d_local.y())
        + c_right.y() * (c_up.z() * hat_d_local.x() - c_up.x() * hat_d_local.z())
        + c_right.x() * (-c_up.z() * hat_d_local.y() + c_up.y() * hat_d_local.z()))
    );
```

Final Step 8: The Local Solar Illumination Angle

The two intercept points of the lines $\mathfrak{L}_1$ and $\mathfrak{L}_2$ and the image plane Ω in rendering coordinates (s_{x1}, s_{y1}) and (s_{x2}, s_{y2}) are the essence after the execution of all previous steps. Finally, the local solar illumination angle α can be derived as an angle between the 2D vectors $(0, -1\,000)^{\mathrm{T}}$ and $(s_{x2} - s_{x1}, s_{y2} - s_{y1})^{\mathrm{T}}$ (see figure 9.4):

$$\mathbf{v}_1 \stackrel{\text{def.}}{=\!=} \begin{pmatrix} 0 \\ -1\,000 \end{pmatrix} \tag{9.21}$$

$$\mathbf{v}_2 \stackrel{\text{def.}}{=\!=} \begin{pmatrix} s_{x2} - s_{x1} \\ s_{y2} - s_{y1} \end{pmatrix} \tag{9.22}$$

$$\alpha = \sphericalangle(\mathbf{v}_1, \mathbf{v}_2) = \begin{cases} \arccos \frac{\langle \mathbf{v}_1, \mathbf{v}_2 \rangle}{\|\mathbf{v}_1\| \|\mathbf{v}_2\|} & \text{for } v_{2,x} \geq 0 \\ 2\pi - \arccos \frac{\langle \mathbf{v}_1, \mathbf{v}_2 \rangle}{\|\mathbf{v}_1\| \|\mathbf{v}_2\|} & \text{otherwise} \end{cases} \tag{9.23}$$

9.3 Determination of the Local Solar Illumination Angle

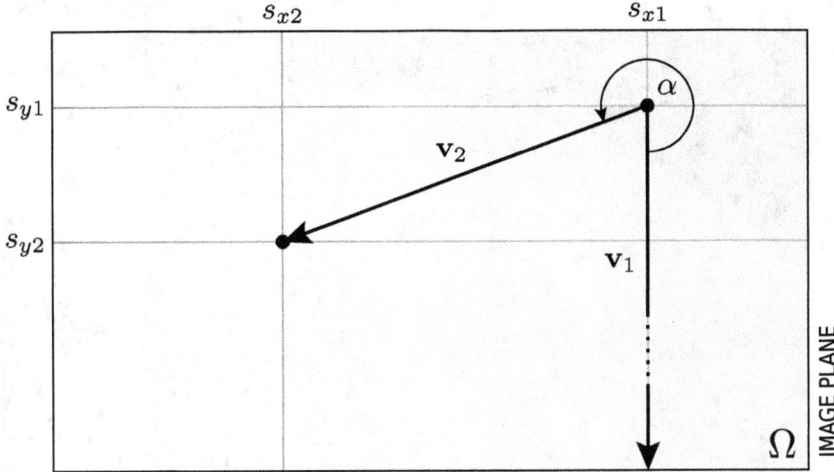

Figure 9.4 Geometrical construction of the local illumination angle α.

```
                        ──── MSIS Implementation ────
1   // STEP 8: The Local Solar Illumination Angle
2   Vector2D v1 = new Vector2D(0, 1000);
3   Vector2D v2 = new Vector2D(x2-x1, y2-y1);
4
5   if (v2.x() > 0)
6   {
7       PixelOut.IlluminationAngle =
8           tools.rad2deg(Math.Acos((v1 * v2) / (v1.norm() * v2.norm())));
9   }
10  else
11  {
12      PixelOut.IlluminationAngle =
13          tools.rad2deg(2 * Math.PI - Math.Acos((v1 * v2) / (v1.norm() * v2.norm())));
14  }
15  PixelOut.exists = true;
16
17  return PixelOut;
```

10 Results, Discussion and Conclusion

10.1 Synopsis of the Work Results

In the course of this thesis work, a software framework for the realistic illumination simulation of the Moon's surface, the *Moon Surface Illumination Simulation Framework* (MSISF) has been developed. The MSISF consists not only of one single application, but rather of a *framework* of software components for distinct tasks. Its main component, the *Moon Surface Illumination Simulator* (MSIS) is the main user interface, which is able to produce the envisaged renderings (an example rendering produced by the MSISF can be seen on the left page).

It has been shown how these renderings will be suitable for the development and testing of new optical navigation algorithms, since not only the renderings themselves are generated, but also these renderings are augmented with general and pixel-wise meta information in the machine-readable XML format. One of the most important items of meta information is the specification of the local solar illumination angle on arbitrary points of the visible Moon's surface on renderings. No other software is currently known to produce these outputs; actual optical navigation algorithms are often tested using example pictures from real space exploration missions, which naturally fail in terms of the availability of the needed ground truth data, whereas the MSIS produces renderings with exactly known parameters for the environmental conditions.

The MSIS also demonstrated its ability not only to produce single renderings, but also whole series of renderings corresponding to a virtual flight trajectory or landing on the Moon at an

---

Chapter Image: MSIS rendering of the Moon at a high altitude.

arbitrary sampling frequency. These rendering series can be assembled into a video of the flight to simulate the continuous video stream of a camera aboard a spacecraft. The MSIS is powerful and — simultaneously — resource-efficient enough to be run on a standard personal computer. In the course of this thesis, videos with durations up to 45 seconds (at 30 frames per second) could be generated on a standard personal computer within 24 hours[1]. Utilizing the power of more than one physical machine, much lower rendering times would be possible. By splitting batch files into separate pieces, the MSIS offers a native way to perform a distributed rendering of rendering series on multiple machines.

At the time of writing this thesis, the DLR Institute of Planetary Research in Berlin-Adlershof also produced a video of the Moon's surface using a DEM, but which was produced by the institute itself using another technique. The LDEMs used in this thesis are produced using laser altimetry, while the DLR Institute of Space Research used a 3D extraction algorithm for stereo images of the Moon's surface. This technique is more demanding than the use of laser altimeters, but it is believed to produce DEMs in far higher resolutions, since every pixel of a stereo image contains information of the topography; this way, the measurement density is much greater than the single laser beam samples of a laser altimeter. However, two weeks and a cluster of 40 computers were needed to produce this DEM and the associated video. [85]

NASA also prepared a video of the Moon from the LRO LOLA data, but at a very high altitude and from a fixed position, showing the lunar phases as seen from Earth for the entire year 2012. NASA states that "[...] using the LRO altimeter data, it can calculate the lengths, directions, and positions of all the shadows of mountains, crater rims, and so on, knowing the angle of the Sun over the horizon" [101].

The MSISF is capable of producing both of the aforementioned videos, too. Even an automatic color-coding of the surface elevation as seen as in [85] or a texture overlay as in [101] would be conceivable using POV-Ray's possibilities.

Due to the separation of the import and conditioning of *digital elevation models* (DEMs) from the MSIS using PHP scripts and a MySQL database with spatial extensions, the MSISF is very flexible from a broader view. The MSISF is not only limited to the Moon; in fact the MSISF is able to produce renderings of all spherical celestial bodies with solid surfaces, for which sufficient DEMs exist.

Utilizing the MSISF, a powerful tool has been developed for the rendering of realistic illuminated solid planetary surfaces, which is able to produce reusable, machine-readable meta information for the purpose of the development and testing of new optical navigation algorithms and systems.

---

[1] These videos are available at http://go.rene-schwarz.com/masters-thesis.

At the very end of this thesis, a short overview of the currently available scientific references in the research field of the illuminance flow estimation on pictures, like they are produced by the MSISF, should have been given. Illuminance flow estimation is expected to give an estimation of the local solar illumination angle for a particular pixel of an image showing a part of a celestial surface. The output of an illuminance flow estimation should ideally be identical to the local solar illumination angle output of the MSIS, assuming that the same definition for the local solar illumination angle is being used. While the MSIS can exactly calculate the local solar illumination angle for a particular pixel out of its information on the scene geometry, an illuminance flow estimation algorithm has to estimate these angles without any additional information on the scene geometry, as is always the case in a real space situation. Unfortunately, there has not been enough time in the narrow timetable of this thesis to get an in-depth look at the proposed algorithms and techniques. However, the presented reference collection (see bibliography in the backmatter) will be a good entry point for further theses and research on this topic.

10.2 Suggestions for Improvements and Future Work

Although the MSISF has been designed and developed very carefully, there are suggestions for starting points of further improvements and additional work, which have not been possible for reasons of time or lack of sufficient data or new software versions.

10.2.1 Performance Optimization

As said before in the introduction of the MSISF in chapter 2, the overall performance of the LDEM import and conditioning process as well as the pattern generation process could be significantly increased by using an optimized storage solution (e.g. solid-state disks in a RAID 0 configuration), if the utilization of LDEM resolutions greater than 64 px/deg is necessary.

An additional starting point for a performance optimization of the rendering process is the open-source POV-Ray rendering software itself. By far the most time is consumed during the rendering process at parsing the surface meshes by POV-Ray, which takes up nearly all the time needed for a rendering. Instead of storing the meshes in a text format, "binary meshes", which means meshes that are stored directly as POV-Ray memory maps, could be used, canceling the requirement for a parsing process. Unfortunately, the POV-Ray team currently has no plans to do this [104]. Nevertheless, POV-Ray is an open-source software; the development of an original renderer, implementing binary meshes, is conceivable.

10.2.2 Improvement of the Topography Database

As soon as it is available, the pattern repository should be rebuilt with NASA LRO LDEMs without interpolated data points. Currently, version 1.05 of the LOLA LDEM data is used, but at the time of completion of this thesis, version 1.09 is available (cf. chapter 4). The usage of DEMs with less interpolated data will significantly increase the quality of the rendered surfaces. Particularly for the polar regions of the Moon, NASA also offers DEMs in polar stereographic projection. An investigation and experimental integration of these data sets into the MSISF could be reasonable.

10.2.3 Graphical User Interface

The development of an additional, graphical user interface (GUI) with the possibility of a 3D preview of the specified simulation timepoints, for example, using the *Windows Presentation Foundation 3D* (WPF 3D), would allow a fast preview and evaluation of the chosen simulation settings. Since the MSIS can be invoked using a normal process call, no changes to the MSISF are necessary; the GUI would simply be added as another tool to the MSISF. Such a GUI would also lower the obstacles for new users.

10.2.4 Rendering Parallelization/Distributed Rendering

Although the MSIS offers a native way of performing a distributed rendering by splitting batch files into several pieces, there should be a solution for the centralized controlling of the rendering process, what could maybe achieved with a client-server solution. Actually, the rendering process on each single machine to be used has to be initiated manually. Additionally, there should be a user option (an additional command-line argument), which allows the user to choose their own file name pattern for the generated output files, simplifying the later assembly process.

10.2.5 Utilization of the MSISF for Other Celestial Bodies

Since NASA science data is usually distributed in file formats that conform to the PDS standard, a rewrite of the import script (ldem_mysql_import.php), which can parse the corresponding PDS label files and cope with the given parameters without the need for manual setting of these parameters could be advantageous in terms of the import of DEMs of other celestial bodies.

10.2.6 Real-Time Video Preparation

With the development of the aforementioned GUI, a real-time video preparation would be possible, because the GUI is only required to display a simplified 3D model as preview, since an ideal sphere is sufficient for preview purposes. The visualization of such a simple 3D model is possible at high sample rates on today's personal computers. The calculation of the light source position utilizing SPICE would be possible at high sample rates, too. This way, the user-given simulation timepoints could be visualized within the GUI at a user-given frame rate. This approach enables the user to have a preview during the preparation of the rendering series, without the need to render each single picture, what could be time saving.

10.2.7 Compile POV-Ray as Windows Command-Line Tool

The currently available Windows version of POV-Ray opens its GUI every time POV-Ray has been invoked (even from the command line). There is no possibility to disable the GUI in Windows. A true batch mode (a command-line-only version of POV-Ray for Windows) will probably be released with version 3.7.0 [103]. If there is no command-line-only version with the release of version 3.7.0, an alternative would be an original compilation, removing the GUI components from the source code.

10.2.8 INI Settings

Constant settings, which will be needed for every MSIS call (e.g. the path to the POV-Ray binary, the output directory, etc.) should be stored in a configuration file and loaded every time the MSIS is executed to achieve maximum comfort for the end-user.

10.2.9 Original Implementation of the 2D Delaunay Triangulation

For time reasons, the 2D-Delaunay triangulation for the surface pattern generation script (generate_pattern.php) has been developed as an external application, which is being called by the pattern generation script. This application has been compiled from a simple MATLAB script (delaunay2D.m, available in the /src/ directory of the MSISF installation path) using the MATLAB Compiler:

```
mcc -o delaunay2D.exe -m delaunay2D.m
```

The delaunay2D.m script only consists of a few lines of code (see section 5.2). The resulting application delaunay2D.exe requires the MATLAB Compiler Runtime (MCR) to be installed on the target system. The MCR is a commercial product, which is not freely available; a distribution of the MCR along with the MSISF is not allowed for license reasons. Additionally, the startup of MATLAB-compiled application takes a long time, probably because of the initialization of the MCR. It is advisable to do an original implementation of the 2D DELAUNAY triangulation or to use available free implementations (such as those from the CGAL, for example) for speed improvements and availability.

10.2.10 Compensate the Drawbacks of the DSPSA

One possibility for the partial compensation of the drawbacks of the *Dynamical Surface Pattern Selection Algorithm* (DSPSA) (cf. section 8.4), which is responsible for the selection of needed surface patterns for a rendering, could be the introduction of a security margin along the periphery of the visible surface area. This would lower the influence of surface features, which are not visible on the later renderings themselves, but which can be detected by their shadows. This solution requires an extension of the DSPSA.

10.3 Construction Progress of TRON

Unfortunately, the construction progress of the TRON facility at the DLR Institute of Space Systems in Bremen has been delayed. At the time of the completion of this thesis, no test image material exists for a comparison of the TRON results with MSIS renderings. Currently, TRON is at the end of the building phase; the calibration of all actors is being prepared at this time. Figure 10.1 shows the current construction state.

However, the TRON development is far advanced: The 6 DOF industrial robot, which will carry the camera and other optical instruments later, as well as the 5 DOF lighting system, are installed and can be controlled in real time using dSPACE real time hardware. The calibration of both systems is carried out with a laser-based tracking system, which is able to localize the positions of all actuators within the TRON facility at an accuracy of ± 0.01 mm. The ability of an accurate positioning of all actuators is an indispensable requirement for the operation of TRON, since the robot's own telemetry shows errors in the magnitude of a couple of millimeters or a tenth of a degree. The positioning of all actuators with an error margin of 1 mm has been a design parameter of the TRON laboratory. Systematic errors in the robot's telemetry are introduced by physical circumstances of the laboratory room itself (uneven walls) as well as fabrication-caused variations in the robot rail systems. Additional research is necessary regarding the occurrence of dynamical effects contributing to the aforementioned overall error, for example, temperature variations or displacement caused by the infinitesimal movement of the surrounding building, vibrations, and so on.

3D surface tiles of the Moon's surface are currently being prepared for installation on the laboratory walls for a project on behalf of the European Space Agency (ESA). An example 3D surface tile is depicted in figure 10.2.

Figure 10.1 Current construction progress of the TRON facility. Visible is the floor-mounted 6 DOF industrial robot simulating a spacecraft as well as the ceiling-mounted 5 DOF lighting system. © DLR; picture reproduced with friendly permission.

10 Results, Discussion and Conclusion

Figure 10.2 A TRON 3D surface tile of the Moon's surface for a project on behalf of the European Space Agency (ESA). © DLR; picture reproduced with friendly permission.

Current plans are for TRON to reach test readiness before June 2012; first test image material is expected a short time after this date.

10.4 MSISF Application at DLR

In the current process of planning, the MSISF will be used for the verification and evaluation of the quality of images generated by TRON once it is ready. Additionally, the MSISF will deliver test images for the testing and evaluation of new spacecraft navigation algorithms, whereas the knowledge and availability of machine-readable, pixel-wise information regarding the local solar illumination angle will be an important advantage over all other existing solutions. With increasing LDEM resolutions, the MSISF is also able to produce ground truth data with unprecedented detail.

In comparison with TRON, the MSISF has several advantages:

1. The MSISF is able to produce realistically curved terrains, which is only possible using a small (in the order of magnitude of 1×1 m) 3D surface tile with TRON, but not over the entire laboratory area.

2. The MSISF can produce virtual, infinitely long, continuous flight trajectories, while TRON is limited to the installed 3D surface tiles and the spatial limitations of the laboratory.

3. A series of renderings can be produced at virtually every sample rate, while TRON is limited to the camera's maximum sample rate. In addition, TRON is only able to produce pictures in a certain, minimum step size because of the limitations introduced by the robot's positioning accuracy.

4. The possibilities for the variation of simulation parameters with reference to flight altitude, illumination angle, field of view, etc. are manifold, whereas the possibilities of TRON are limited by the physical properties of the camera or quality of the fabricated 3D surface tiles (terrain models) to be used.

5. The simulation parameters of the MSISF are exactly known.

Nevertheless, TRON also has advantages compared with the MSISF:

1. Only a physical simulation allows a direct qualification of space hardware using real-time hardware-in-the-loop (HiL) configurations.

2. Generally, a qualification of other space hardware (e.g. combined camera navigation systems) is possible in the first place, using a physical experimental assembly.

3. Only the usage of real cameras allows an evaluation of the influences of optical effects, which are difficult to model using a software system (or impossible, if real time conditions are required), on navigation algorithms.

4. TRON allows a physically accessible simulation of the application of hardware in space scenarios.

However, only both components in conjunction will offer a holistic solution for the development and testing of new optical navigation systems. Both MSISF and TRON together will have the potential to make a significant contribution to research for the next generation of space exploration systems.

MSIS User Interface Specification

— table beginning at the next page —

A MSIS User Interface Specification

Option	Description	Shorthd.
`--time MJD`	Simulation time as Modified Julian Date (MJD(UTC))	-t
`--times {MJD1,MJD2,MJD3,...}`	List of discrete time steps for the simulation time as Modified Julian Dates (MJD(UTC)).	-tt
`--time-interval [start_time:stepsize:end_time]`	An interval of discrete time steps for the simulation time as Modified Julian Dates (MJD); specified using a start time, a stepsize and end time. The single points of time will be calculated using the given stepsize (MATLAB syntax). REMARK: Be careful with this option, since it can produce a large amount of renderings.	
`--epoch MJD`	Point of time (epoch) as Modified Julian Date MJD(UTC) for whom the given orbital parameters are valid.	-e

`--kepler-set {a,e,omega,Omega,i,M0}`	A traditional set of KEPLERian Orbit Elements for the S/C orbit. Consists of
	    a - semi-major axis [m]
	    e - eccentricity [1]
	    omega - argument of periapsis [rad]
	    Omega - longitude of ascending node (LAN) [rad]
	    i - inclination [rad]
	    M0 - mean anomaly at epoch [rad]
	This option must not be used, if a set of cartesian state vectors (-s, --state-vectors) has been given before.
		-k

--state-vectors {rx,ry,rz,drx,dry,drz} -s	Cartesian state vectors specifying the S/C orbit. Consists of rx - position of the S/C w.r.t. Moon's ME/PA reference frame (x-axis) [m] ry - position of the S/C w.r.t. Moon's ME/PA reference frame (y-axis) [m] rz - position of the S/C w.r.t. Moon's ME/PA reference frame (z-axis) [m] drx - velocity of the S/C w.r.t. Moon's ME/PA reference frame (x-axis) [m/s] dry - velocity of the S/C w.r.t. Moon's ME/PA reference frame (y-axis) [m/s] drz - velocity of the S/C w.r.t. Moon's ME/PA reference frame (z-axis) [m/s] This option must not be used, if a set of KEPLERian orbit elements (-k, --kepler-set) has been given before.
--pattern-repos PATH -p	Set the path of the pattern repository. DEFAULT: ./pattern-repository/
--pov-path PATH	Set the path of the POV-Ray executable (pvengine64.exe). DEFAULT: ./POV-Ray-3.7/bin/
--output-dir PATH -o	Directory for the output files. DEFAULT: ./output/
--fov DEGREES -f	Adjusts the camera field of view (FOV). Must be given in degrees. DEFAULT: 40
--attitude {qx,qy,qz,q0} -a	Sets the S/C orientation according to a given quaternion $q = q0 + qx\ i + qy\ j + qz\ k$ at epoch. Without this option, the S/C is pointed in the direction of Nadir, aligned to the north pole by default.

A MSIS User Interface Specification

--attitude-transition {dpx,dpy,dpz}	Set the S/C orientation transition w.r.t. time by a vector consisting of angular velocities around the respective axis of the S/C. DEFAULT: {0,0,0}	-at
--width WIDTH	Sets the width in pixel of the rendered image(s). DEFAULT: 1024	-w
--height HEIGHT	Sets the height in pixel of the rendered image(s). DEFAULT: 1024	-h
--res RESOLUTION	Sets the resolution of the used LOLA LDEM patterns for the rendering. DEFAULT: 4	-r
--batch	Close application after finishing operations. This option should be used in batch mode or when MoonIllumSim.exe is called from another process.	-b
--ignore-sun	Instructs the MSIS to ignore the Sun's position; the light source will be placed directly at the camera position. Useful if no MJD is known, at whom the Sun will illuminate the scene.	
--rendering-annotation	Produce an additional rendering with auxiliary information (simulation timecode, s/c position and orientation, Sun position, flight altitude, surface mesh resolution, FOV etc.) and a visualization of the local solar illumination angles.	
--fixed-state {MJD,rx,ry,rz,qx,qy,qz,q0}	Disable orbit calculations; use a fixed s/c position and orientation at a given time instead.	

`--batch-file FILENAME`	Disable orbit calculations; use fixed s/c positions and orientations at given discrete times. BATCH FILE CONTENTS: The batch file consists of one line for each s/c state; each line has to satisfy the following pattern (where whitespaces between the single parts represent an arbitrary amount of tabulators or whitespaces): `MJD rx ry rz qx qy qz q0`	
`--grid SPACING`	Sets the grid spacing (both horizontally/vertically) for the output of the meta information (illumination direction, latitude/longitude etc.). SPACING is an unsigned integer value >0, giving the horizontal and vertical spacing in pixels. DEFAULT: 10	`-g`
`--gridH SPACING`	Sets the grid horizontal spacing for the output of the meta information (illumination direction, latitude/longitude etc.). SPACING is an unsigned integer value >0, giving the horizontal spacing in pixels. DEFAULT: 10	`-gH`
`--gridV SPACING`	Sets the grid vertical spacing for the output of the meta information (illumination direction, latitude/longitude etc.). SPACING is an unsigned integer value >0, giving the vertical spacing in pixels. DEFAULT: 10	`-gV`

APPENDIX B

Code Listings

B.1 MySQL Server Instance Configuration

```
 1  # MySQL Server Instance Configuration File
 2  # ----------------------------------------------------------------------
 3  # Generated by the MySQL Server Instance Configuration Wizard
 4  #
 5  #
 6  # Installation Instructions
 7  # ----------------------------------------------------------------------
 8  #
 9  # On Linux you can copy this file to /etc/my.cnf to set global options,
10  # mysql-data-dir/my.cnf to set server-specific options
11  # (@localstatedir@ for this installation) or to
12  # ~/.my.cnf to set user-specific options.
13  #
14  # On Windows you should keep this file in the installation directory
15  # of your server (e.g. C:\Program Files\MySQL\MySQL Server X.Y). To
16  # make sure the server reads the config file use the startup option
17  # "--defaults-file".
18  #
19  # To run run the server from the command line, execute this in a
20  # command line shell, e.g.
21  # mysqld --defaults-file="C:\Program Files\MySQL\MySQL Server X.Y\my.ini"
22  #
23  # To install the server as a Windows service manually, execute this in a
24  # command line shell, e.g.
25  # mysqld --install MySQLXY --defaults-file="C:\Program Files\MySQL\MySQL Server X.Y\my.ini"
```

```
#
# And then execute this in a command line shell to start the server, e.g.
# net start MySQLXY
#
#
# Guildlines for editing this file
# -----------------------------------------------------------------------
#
# In this file, you can use all long options that the program supports.
# If you want to know the options a program supports, start the program
# with the "--help" option.
#
# More detailed information about the individual options can also be
# found in the manual.
#
#
# CLIENT SECTION
# -----------------------------------------------------------------------
#
# The following options will be read by MySQL client applications.
# Note that only client applications shipped by MySQL are guaranteed
# to read this section. If you want your own MySQL client program to
# honor these values, you need to specify it as an option during the
# MySQL client library initialization.
#
[client]
port=3306
#ssl-ca="C:/srv/mysql/ca-cert.pem"
#ssl-cert="C:/srv/mysql/client-cert.pem"
#ssl-key="C:/srv/mysql/client-key.pem"

[mysql]

default-character-set=utf8


# SERVER SECTION
# -----------------------------------------------------------------------
#
# The following options will be read by the MySQL Server. Make sure that
# you have installed the server correctly (see above) so it reads this
# file.
#
[mysqld]
# The TCP/IP Port the MySQL Server will listen on
port=3306
#ssl-ca="C:/srv/mysql/ca-cert.pem"
```

B.1 MySQL Server Instance Configuration

```
#ssl-cert="C:/srv/mysql/server-cert.pem"
#ssl-key="C:/srv/mysql/server-key.pem"
#server-id=1
#log-bin="C:/srv/mysql/replication/mysql-bin.bin"

max_allowed_packet=16M


#Path to installation directory. All paths are usually resolved relative to this.
basedir="S:/srv/Software/mysql-5.5.20-winx64/"

#Path to the database root
datadir="S:/srv/mysql/data/"
log-error="S:/srv/mysql/logs/error.log"

# The default character set that will be used when a new schema or table is
# created and no character set is defined
character-set-server=utf8

# The default storage engine that will be used when create new tables when
default-storage-engine=MYISAM

# Set the SQL mode to strict
sql-mode="STRICT_TRANS_TABLES,NO_AUTO_CREATE_USER,NO_ENGINE_SUBSTITUTION"

# The maximum amount of concurrent sessions the MySQL server will
# allow. One of these connections will be reserved for a user with
# SUPER privileges to allow the administrator to login even if the
# connection limit has been reached.
max_connections=800

# Query cache is used to cache SELECT results and later return them
# without actual executing the same query once again. Having the query
# cache enabled may result in significant speed improvements, if your
# have a lot of identical queries and rarely changing tables. See the
# "Qcache_lowmem_prunes" status variable to check if the current value
# is high enough for your load.
# Note: In case your tables change very often or if your queries are
# textually different every time, the query cache may result in a
# slowdown instead of a performance improvement.
query_cache_size=84M

# The number of open tables for all threads. Increasing this value
# increases the number of file descriptors that mysqld requires.
# Therefore you have to make sure to set the amount of open files
# allowed to at least 4096 in the variable "open-files-limit" in
# section [mysqld_safe]
```

```
table_cache=1520

# Maximum size for internal (in-memory) temporary tables. If a table
# grows larger than this value, it is automatically converted to disk
# based table This limitation is for a single table. There can be many
# of them.
tmp_table_size=1G


# How many threads we should keep in a cache for reuse. When a client
# disconnects, the client's threads are put in the cache if there aren't
# more than thread_cache_size threads from before.  This greatly reduces
# the amount of thread creations needed if you have a lot of new
# connections. (Normally this doesn't give a notable performance
# improvement if you have a good thread implementation.)
thread_cache_size=38

#*** MyISAM Specific options

# The maximum size of the temporary file MySQL is allowed to use while
# recreating the index (during REPAIR, ALTER TABLE or LOAD DATA INFILE.
# If the file-size would be bigger than this, the index will be created
# through the key cache (which is slower).
myisam_max_sort_file_size=100G

# If the temporary file used for fast index creation would be bigger
# than using the key cache by the amount specified here, then prefer the
# key cache method.  This is mainly used to force long character keys in
# large tables to use the slower key cache method to create the index.
myisam_sort_buffer_size=30M

# Size of the Key Buffer, used to cache index blocks for MyISAM tables.
# Do not set it larger than 30% of your available memory, as some memory
# is also required by the OS to cache rows. Even if you're not using
# MyISAM tables, you should still set it to 8-64M as it will also be
# used for internal temporary disk tables.
key_buffer_size=5G

# Size of the buffer used for doing full table scans of MyISAM tables.
# Allocated per thread, if a full scan is needed.
read_buffer_size=250M
read_rnd_buffer_size=500M

# This buffer is allocated when MySQL needs to rebuild the index in
# REPAIR, OPTIMZE, ALTER table statements as well as in LOAD DATA INFILE
# into an empty table. It is allocated per thread so be careful with
# large settings.
```

```
sort_buffer_size=500M


#*** INNODB Specific options ***
innodb_data_home_dir="S:/srv/mysql/data/InnoDB/"

# Use this option if you have a MySQL server with InnoDB support enabled
# but you do not plan to use it. This will save memory and disk space
# and speed up some things.
#skip-innodb

# Additional memory pool that is used by InnoDB to store metadata
# information.  If InnoDB requires more memory for this purpose it will
# start to allocate it from the OS.  As this is fast enough on most
# recent operating systems, you normally do not need to change this
# value. SHOW INNODB STATUS will display the current amount used.
innodb_additional_mem_pool_size=6M

# If set to 1, InnoDB will flush (fsync) the transaction logs to the
# disk at each commit, which offers full ACID behavior. If you are
# willing to compromise this safety, and you are running small
# transactions, you may set this to 0 or 2 to reduce disk I/O to the
# logs. Value 0 means that the log is only written to the log file and
# the log file flushed to disk approximately once per second. Value 2
# means the log is written to the log file at each commit, but the log
# file is only flushed to disk approximately once per second.
innodb_flush_log_at_trx_commit=1

# The size of the buffer InnoDB uses for buffering log data. As soon as
# it is full, InnoDB will have to flush it to disk. As it is flushed
# once per second anyway, it does not make sense to have it very large
# (even with long transactions).
innodb_log_buffer_size=3M

# InnoDB, unlike MyISAM, uses a buffer pool to cache both indexes and
# row data. The bigger you set this the less disk I/O is needed to
# access data in tables. On a dedicated database server you may set this
# parameter up to 80% of the machine physical memory size. Do not set it
# too large, though, because competition of the physical memory may
# cause paging in the operating system.  Note that on 32bit systems you
# might be limited to 2-3.5G of user level memory per process, so do not
# set it too high.
innodb_buffer_pool_size=250M

# Size of each log file in a log group. You should set the combined size
# of log files to about 25%-100% of your buffer pool size to avoid
# unneeded buffer pool flush activity on log file overwrite. However,
```

```
# note that a larger logfile size will increase the time needed for the
# recovery process.
innodb_log_file_size=50M

# Number of threads allowed inside the InnoDB kernel. The optimal value
# depends highly on the application, hardware as well as the OS
# scheduler properties. A too high value may lead to thread thrashing.
innodb_thread_concurrency=50
```

B.2 MSISRendering XML Document Type Definition (DTD)

```
<!--
##########################################################################
I N F O H E A D E R

N O T E !
This DTD file belongs to any MoonSurfIllumSim_step_[vwxyz].xml file and
it should be wrapped in a DOCTYPE definition with the following syntax:
<!DOCTYPE root-element SYSTEM "filename">

E L E M E N T   I N C I D E N C E   D E C L A R A T I O N
? - none / once
+ - once to infinite
* - none to infinite
<!ELEMENT element-name (child-name?,child-name+,child-name*)

D E C L A R I N G   E I T H E R / O R   C O N T E N T
Example: <!ELEMENT note (to,from,header,(message|body))>

A T T R I B U T E   V A L U E S
PLEASE CHECK URL: http://www.w3schools.com/dtd/dtd_attributes.asp

E N D   O F   I N F O   H E A D E R
##########################################################################
-->

<!-- R O O T   E L E M E N T   A N D   C H I L D   E L E M E N T S
     I N   M A N D A N T O R Y   O R D E R -->
<!ELEMENT MSISRendering (GeneralInformation, PixelInformation)>

<!-- C H I L D   E L E M E N T   #1   A N D   E X T E N D I N G
     E L E M E N T S   I N   M A N D A N T O R Y   O R D E R -->
<!ELEMENT GeneralInformation (SimulationTime,CameraPosition,CameraOrientation,SunPosition,▼
```

B.2 MSISRendering XML Document Type Definition (DTD)

```dtd
FlightAltitude,SurfaceResolution,FOV,MSISVersion,CommandLine)>

<!-- S I M U L A T I O N   T I M E   A N D   E X T E N D I N G   E L E M E N T S -->
<!ELEMENT SimulationTime (MJD,UTC)>
<!ELEMENT MJD (#PCDATA)>
<!ELEMENT UTC (#PCDATA)>

<!-- C A M E R A   P O S I T I O N -->
<!ELEMENT CameraPosition (Vector3D)>
<!ELEMENT Vector3D EMPTY>
<!ATTLIST CameraPosition unit (m) #FIXED "m">
<!ATTLIST Vector3D x CDATA #REQUIRED y CDATA #REQUIRED z CDATA #REQUIRED>

<!-- C A M E R A   O R I E N T A T I O N -->
<!ELEMENT CameraOrientation (Quaternion)>
<!ELEMENT Quaternion EMPTY>
<!ATTLIST Quaternion r CDATA #REQUIRED x CDATA #REQUIRED y CDATA #REQUIRED z
CDATA #REQUIRED>

<!-- S U N   P O S I T I O N -->
<!ELEMENT SunPosition (Vector3D)>
<!ATTLIST SunPosition unit (m) #FIXED "m">
<!ATTLIST Vector3D x CDATA #REQUIRED y CDATA #REQUIRED z CDATA #REQUIRED>

<!-- F L I G H T   A L T I T U D E -->
<!ELEMENT FlightAltitude (#PCDATA)>
<!ATTLIST FlightAltitude unit (m) #FIXED "m">

<!-- S U R F A C E   R E S O L U T I O N -->
<!ELEMENT SurfaceResolution (#PCDATA)>
<!ATTLIST SurfaceResolution unit CDATA "px/deg">

<!-- F O V -->
<!ELEMENT FOV (#PCDATA)>
<!ATTLIST FOV unit (deg) #FIXED "deg">

<!-- M S I S   V E R S I O N -->
<!ELEMENT MSISVersion (#PCDATA)>

<!-- COMMAND LINE ARGUMENTS -->
<!ELEMENT CommandLine (#PCDATA)>

<!-- C H I L D   E L E M E N T   # 2   A N D   E X T E N D I N G   E L E M E N T -->
<!ELEMENT PixelInformation (Pixel+)>

<!-- P I X E L -->
<!ELEMENT Pixel (SelenographicCoordinates,IlluminationDirection)>
```

```
80  <!ATTLIST Pixel x CDATA #REQUIRED y CDATA #REQUIRED>
81
82  <!ELEMENT SelenographicCoordinates EMPTY>
83  <!ATTLIST SelenographicCoordinates lat CDATA #REQUIRED lon CDATA #REQUIRED units ▼
84  (deg) #FIXED "deg">
85
86  <!ELEMENT IlluminationDirection (#PCDATA)>
87  <!ATTLIST IlluminationDirection unit (deg) #FIXED "deg">
```

B.3 LDEM Import Script

```php
<?php
# *****************************************************************************
#                        _
#                       | |
#       _ __   _ __    _| |__     ___   _  _   _
#      | '_/ _ \ '_ \ / _ \/ _|/ _| '_ \\/\/ / _'| '_|_ /
#      | | |  _/| | |  _/ \_ \ (_| | | | \ v v / (_| | | / /
#      |_|  \__|_| |_|\__| |__/\__|_| |_|\_/\_/ \__,_|_| /__|
#                                                   rene-schwarz.com
#
# *****************************************************************************
#       MSISF LRO LOLA LUNAR DIGITAL ELEVATION MODEL MYSQL IMPORT SCRIPT
# *****************************************************************************
#
#    Author:         B.Eng. René Schwarz
#                    mail: <mail@rene-schwarz.com>
#                    web:  http://www.rene-schwarz.com
#                    on behalf of the German Aerospace Center (DLR)
#    Date:           2012/01/30
#    Filename:       ldem_mysql_insert.php
#    Version:        1.0
#    License:        GNU General Public License (GPL), version 2 or later
#
# *****************************************************************************
/**
 * @mainpage MSISF LRO LOLA Lunar Digital Elevation Model MySQL Import Script
 * @brief
 *    MSISF LRO LOLA Lunar Digital Elevation Model MySQL Import Script
 *
 *    This script imports NASA LRO LOLA LDEM datasets into a spatial MySQL
 *    database. This database is required by the MSISF for the generation of
 *    the surface pattern repository.
 *
 *    LDEM import has only to be done once per resolution or after an update of
```

```
 *     NASA files for a certain resolution. Afterwards all surface patterns of this
 *     specific resolution have to be re-generated using the generate_pattern.php
 *     script. If a pattern repository already exists with actual data, neither a
 *     MySQL import nor a (re-)generation of surface patterns is necessary.
 *
 *     Please be aware that the LDEM data import process for resolutions greater
 *     than LDEM_16 can require a considerable amount of time and resources,
 *     depending on your hardware configuration. Import times in the order of
 *     hours are not unusual.
 *
 *     For more information, see my master's thesis,
 *     available at <http://go.rene-schwarz.com/masters-thesis>.
 *
 *
 *     Copyright (c) 2011 B.Eng. René Schwarz <mail@rene-schwarz.com>
 *     Licensed under the GNU General Public License (GPL), version 2 or later
 *
 *
 *
 *     Requirements:
 *     - MySQL database with tables prepared for the lunar topographic data as
 *       specified in [1]
 *     - PHP >= version 5.3.9
 *     - corresponding NASA LRO LOLA LDEM files, available from
 *       <http://imbrium.mit.edu/DATA/LOLA_GDR/CYLINDRICAL/IMG/>, which shall be
 *       imported
 *
 *
 *     References:
 *     [1] Schwarz, René: Development of an illumination simulation software for
 *         the Moon's surface: An approach to illumination direction estimation
 *         on pictures of planetary solid surfaces with a significant number of
 *         craters. Master's Thesis, Merseburg University of Applied Sciences,
 *         German Aerospace Center (DLR). Books on Demand, Norderstedt, Germany,
 *         2012.
 *
 *
 * @author B.Eng. René Schwarz <mail@rene-schwarz.com>
 * @version 1.0
 * @date 2012/01/30
 */
# ========================================================
# Configuration of the MySQL import process
# ========================================================

# LDEM resolution (integer) to be used (preconfigured for LDEM_4, LDEM_16,
# LDEM_64, LDEM_128, LDEM_256, LDEM_512 and LDEM_1024).
```

B Code Listings

```
$LDEM = 1024;

# Path to LDEM directory, where all LDEM files (*.img) to be imported are placed
# (w/o trailing slash)
$path_LDEM = "L:/path/to/LDEM";

# MySQL connection parameters
$mysql_host = "localhost";      # MySQL server, e.g. "localhost" or "localhost:3306"
$mysql_db = "";                 # MySQL database
$mysql_user = "";               # MySQL username for specified database
$mysql_pwd = "";                # MySQL password for specified username

# LDEM-specific options
# These parameters can be found in the corresponding *.lbl files for each *.img file.
# Preconfigured for LDEM_4, LDEM_16, LDEM_64, LDEM_128, LDEM_256, LDEM_512 and LDEM_1024.
switch($LDEM)
{
    case 4:
    default:
    /* LDEM_4 */
        $c_MAP_RESOLUTION = 4;
        $c_LINE_SAMPLES = 1440;
        $c_LINE_LAST_PIXEL = 720;
        $c_LINE_PROJECTION_OFFSET = 359.5;
        $c_SAMPLE_PROJECTION_OFFSET = 719.5;
        break;

    case 16:
    /* LDEM_16 */
        $c_MAP_RESOLUTION = 16;
        $c_LINE_SAMPLES = 5760;
        $c_LINE_LAST_PIXEL = 2880;
        $c_LINE_PROJECTION_OFFSET = 1439.5;
        $c_SAMPLE_PROJECTION_OFFSET = 2879.5;
        break;

    case 64:
    /* LDEM_64 */
        $c_MAP_RESOLUTION = 64;
        $c_LINE_SAMPLES = 23040;
        $c_LINE_LAST_PIXEL = 11520;
        $c_LINE_PROJECTION_OFFSET = 5759.5;
        $c_SAMPLE_PROJECTION_OFFSET = 11519.5;
        break;

    case 128:
    /* LDEM_128 */
```

```
            $c_MAP_RESOLUTION = 128;
            $c_LINE_SAMPLES = 46080;
            $c_LINE_LAST_PIXEL = 23040;
            $c_LINE_PROJECTION_OFFSET = 11519.5;
            $c_SAMPLE_PROJECTION_OFFSET = 23039.5;
            break;

    case 256:
    /* LDEM_256 */
            $c_MAP_RESOLUTION = 256;
            $c_LINE_SAMPLES = 46080;
            $c_LINE_LAST_PIXEL = 23040;
            break;

    case 512:
    /* LDEM_512 */
            $c_MAP_RESOLUTION = 512;
            $c_LINE_SAMPLES = 46080;
            $c_LINE_LAST_PIXEL = 23040;
            break;

    case 1024:
    /* LDEM_1024 */
            $c_MAP_RESOLUTION = 1024;
            $c_LINE_SAMPLES = 30720;
            $c_LINE_LAST_PIXEL = 15360;
            break;
}

# ===========================================================
#                    END CONFIGURATION
#
#              DON'T EDIT BEYOND THIS LINE!
# ===========================================================

$mtime = microtime();
$mtime = explode(" ",$mtime);
$mtime = $mtime[1] + $mtime[0];
$starttime = $mtime;

function cline($str = "")
{
    print("{$str}\r\n");
}

$c_SAMPLE_BITS = 16;
$c_RECORD_BYTES = ($c_LINE_SAMPLES * $c_SAMPLE_BITS)/8;
```

```php
$c_CENTER_LATITUDE = 0;
$c_CENTER_LONGITUDE = 180;
$c_SCALING_FACTOR = 0.5;
$c_OFFSET = 1737400;

if($LDEM > 128)
{
    if(!isset($argv[1]))
    {
        die("For LDEM resolutions greater than 128 px/deg files have been\r\n" .
            "splitted by NASA. Additional command-line arguments are necessary\r\n" .
            "for the specification of the certain LDEM part. The values can\r\n" .
            "be found in the corresponding *.lbl file.\r\n\r\n" .
            "USAGE:\r\n" .
            "php ldem_mysql_insert.php additionalFilenamePart LINE_PROJECTION_OFFSET " .
            "SAMPLE_PROJECTION_OFFSET\r\n\r\n" .
            "EXAMPLE:\r\n" .
            "php ldem_mysql_insert.php 00N_15N_330_360 15359.5 -153600.5\r\n");
    }

    $filename = "{$path_LDEM}/LDEM_{$c_MAP_RESOLUTION}_{$argv[1]}.img";
    $c_LINE_PROJECTION_OFFSET = $argv[2];
    $c_SAMPLE_PROJECTION_OFFSET = $argv[3];
}
else
{
    $filename = "{$path_LDEM}/LDEM_{$c_MAP_RESOLUTION}.img";
}

if($handle = fopen($filename, "r"))
{
    cline("NASA LOLA data file \"{$filename}\" opened.");
}
else
{
    die("NASA LOLA data file \"{$filename}\" could not be opened.");
}

cline("Connecting to MySQL database...");
$mysqli = new mysqli($mysql_host, $mysql_user, $mysql_pwd, $mysql_db);
cline("Connection to MySQL database established.");

for($line = 1; $line <= $c_LINE_LAST_PIXEL; $line++)
{
    $mtime = microtime();
    $mtime = explode(" ",$mtime);
    $mtime = $mtime[1] + $mtime[0];
```

B.3 LDEM Import Script

```php
        $lstarttime = $mtime;

        fseek($handle, ($line - 1) * $c_RECORD_BYTES);
        $line_content = unpack("s*", fread($handle, $c_RECORD_BYTES));

        $sql = "INSERT INTO LDEM_{$c_MAP_RESOLUTION} (lat, lon, planetary_radius, x, y, z, " .
            "point) VALUES ";

        $j = 1;
        for($sample = 1; $sample <= $c_LINE_SAMPLES; $sample++)
        {
            $dn = $line_content[$sample];

            $point["lat"] = $c_CENTER_LATITUDE - ($line - $c_LINE_PROJECTION_OFFSET - 1)
                            / $c_MAP_RESOLUTION;
            $point["lon"] = $c_CENTER_LONGITUDE +
                            ($sample - $c_SAMPLE_PROJECTION_OFFSET - 1) / $c_MAP_RESOLUTION;
            $point["height"] = ($dn * $c_SCALING_FACTOR);
            $point["planetary_radius"] = $point["height"] + $c_OFFSET;
            $point["x"] = $point["planetary_radius"] * cos(deg2rad($point["lat"]))
                            * cos(deg2rad($point["lon"]));
            $point["y"] = $point["planetary_radius"] * cos(deg2rad($point["lat"]))
                            * sin(deg2rad($point["lon"]));
            $point["z"] = $point["planetary_radius"] * sin(deg2rad($point["lat"]));
            $point["point"] = "PointFromText('POINT({$point["lat"]} {$point["lon"]})')";

            if(substr($sql, -7) == "VALUES ")
            {
                $sql .="('{$point["lat"]}', '{$point["lon"]}', '{$point["planetary_radius"]}'".
                    ", '{$point["x"]}', '{$point["y"]}', '{$point["z"]}', {$point["point"]})";
            }
            else
            {
                $sql .=", ('{$point["lat"]}', '{$point["lon"]}', '{$point["planetary_radius"]}'".
                    ", '{$point["x"]}', '{$point["y"]}', '{$point["z"]}', {$point["point"]})";
            }

            $j++;

            // prevent extreme huge SQL queries
            // (cut at 24000 inserts, since one LDEM_64 line contains 23,040 values)
            if($j > 24000)
            {
                if(!$mysqli->query($sql))
                {
                    $mysqli->close();
                    die("Error: " . $mysqli->error);
```

```php
                }
                $sql = "INSERT INTO LDEM_{$c_MAP_RESOLUTION} (lat, lon, planetary_radius, x, y" .
                    ", z, point) VALUES ";
                $j = 1;
            }
        }

        if(!$mysqli->query($sql))
        {
            $mysqli->close();
            die("Error: " . $mysqli->error);
        }

        $mtime = microtime();
        $mtime = explode(" ",$mtime);
        $mtime = $mtime[1] + $mtime[0];
        $lendtime = $mtime;
        $ltotaltime = ($lendtime - $lstarttime);
        cline("Line {$line} of {$c_LINE_LAST_PIXEL} processed (" .
            (memory_get_peak_usage()/(1024*1024)) . " MiB, total execution time:" .
            "{$ltotaltime} sec.)..."
        );
    }

    cline();
    cline();
    $mysqli->close();
    cline("MySQL connection closed.");

    $mtime = microtime();
    $mtime = explode(" ",$mtime);
    $mtime = $mtime[1] + $mtime[0];
    $endtime = $mtime;
    $totaltime = ($endtime - $starttime);
    cline();
    cline("Operation has finished.");
    cline("Peak memory needed: " . (memory_get_peak_usage()/(1024*1024)) . " MiB, total" .
        "execution time: {$totaltime} sec."
    );

?>
```

B.4 Pattern Generation Script

```php
<?php
# *****************************************************************************
#                             _
#                            | |
#     __ _ _ __   __      __ | |__      __      _____  __
#    |  _/ _ \ _ \ / _ \ /  _|/ _| '_\ \  /\  / / _` | '_|_ /
#    | | |  _/| | |  _/ \_ \ (_| | | | \ v v / (_| | | |  / /
#    |_|  \___|_| |_|\___|  __/\__|_| |_|\_/\_/ \__,_|_| /___|
#                                                        rene-schwarz.com
#
# *****************************************************************************
#                   MSISF SURFACE PATTERN GENERATION SCRIPT
# *****************************************************************************
#
#   Author:         B.Eng. René Schwarz
#                     mail: <mail@rene-schwarz.com>
#                     web:  http://www.rene-schwarz.com
#                   on behalf of the German Aerospace Center (DLR)
#   Date:           2012/01/11
#   Filename:       generate_pattern.php
#   Version:        1.0
#   License:        GNU General Public License (GPL), version 2 or later
#
# *****************************************************************************
/**
 * @mainpage MSISF Surface Pattern Generation Script
 * @brief
 *   MSISF Surface Pattern Generation Script
 *
 *   This script generates the MSISF-required surface patterns out of a spatial
 *   MySQL database for the Moon's surface (made out of NASA LRO LOLA LDEM files).
 *
 *   Pattern generation have only to be done once per resolution or after an
 *   modification of the database data for a certain resolution, since all
 *   generated patterns will be placed in a central pattern repository, being
 *   available for the MSISF.
 *
 *   Please be aware that the pattern generation process for resolutions greater
 *   than LDEM_16 can require a considerable amount of time and resources,
 *   depending on your hardware configuration. Generation times in the order of
 *   weeks are not unusual.
 *
 *   For more information, see my master's thesis,
 *   available at <http://go.rene-schwarz.com/masters-thesis>.
```

```php
 *
 *
 *     Copyright (c) 2011 B.Eng. René Schwarz <mail@rene-schwarz.com>
 *     Licensed under the GNU General Public License (GPL), version 2 or later
 *
 *
 *
 *     Requirements:
 *     - MySQL database loaded with lunar topographic data as specified in [1]
 *     - delaunay2D.exe, as distributed with this script
 *     - Matlab R2011b or MATLAB Compiler Runtime (MCR) v7.16
 *     - PHP >= version 5.3.9
 *
 *
 *     References:
 *     [1] Schwarz, René: Development of an illumination simulation software for
 *         the Moon's surface: An approach to illumination direction estimation
 *         on pictures of planetary solid surfaces with a significant number of
 *         craters. Master's Thesis, Merseburg University of Applied Sciences,
 *         German Aerospace Center (DLR). Books on Demand, Norderstedt, Germany,
 *         2012.
 *
 *
 * @author B.Eng. René Schwarz <mail@rene-schwarz.com>
 * @version 1.0
 * @date 2012/01/11
 */


# Raise PHP's memory limit to 1 GB of RAM to ensure data handling (enough up to
# LDEM_64). If you are trying to import LDEMs greater than 64 px/deg in resolution,
# increase the memory limit by testing.
ini_set('memory_limit', '1G');


# ========================================================
# Configuration of the surface pattern generation process
# ========================================================

# LDEM resolution (integer) to be used (preconfigured for LDEM_4, LDEM_16 and LDEM_64).
$LDEM = 64;

# Path to pattern repository, where all generated patterns will be placed and where
# all existing patterns are located (w/o trailing slash)
$path_patternDB = "L:/path/to/pattern-repository";

# Path to a temporary directory (w/o trailing slash)
```

```
$path_tempdir = "L:/path/to/temp-dir";

# Path to the supplied delaunay2D.exe file (w/o trailing slash)
$path_delaunayHelper = "L:/path/to/MSISF-installation/bin";

# MySQL connection parameters
$mysql_host = "localhost";      # MySQL server, e.g. "localhost" or "localhost:3306"
$mysql_db   = "";               # MySQL database
$mysql_user = "";               # MySQL username for specified database
$mysql_pwd  = "";               # MySQL password for specified username

# Surface pattern offset (preconfigured for LDEM_4, LDEM_16 and LDEM_64)
# This offset specifies the latitude and longitude, which will be added to 5°x5° as
# overlap area between the single surface patterns to ensure a closed 3D surface during
# rendering/raytracing. For resolutions greater than 64 px/deg, 0.1° should be sufficient.
# All values of $off are given in degrees latitude/longitude.
switch($LDEM)
{
    case 4:
        $off = 0.5;
        break;
    case 16:
        $off = 0.5;
        break;
    case 64:
        $off = 0.1;
        break;
    default:
        die("No valid LDEM dataset selected.");
        break;
}

# ===========================================================
#                      END CONFIGURATION
#
#              DON'T EDIT BEYOND THIS LINE!
# ===========================================================


function cline($str = "")
{
    print("{$str}\r\n");
}


$mtime = microtime();
```

```
139    $mtime = explode(" ",$mtime);
140    $mtime = $mtime[1] + $mtime[0];
141    $starttime = $mtime;
142
143    cline("Connecting to MySQL database...");
144    $db_link = mysql_connect($mysql_host, $mysql_user, $mysql_pwd)
145            OR die(mysql_error($db_link));
146    mysql_select_db($mysql_db, $db_link) OR die(mysql_error($db_link));
147    cline("Connection to MySQL database established.");
148
149    for($lat = -90; $lat <= 85; $lat += 5)
150    {
151        $lat_s = $lat;
152        $lat_e = $lat + 5;
153        $lat_start = $lat - $off;
154        $lat_end = $lat + 5 + $off;
155
156        for($lon = 0; $lon <= 355; $lon += 5)
157        {
158            $mtime = explode(" ",microtime());
159            $mtime = $mtime[1] + $mtime[0];
160            $starttime1 = $mtime;
161
162            $lon_s = $lon;
163            $lon_e = $lon + 5;
164            $lon_start = $lon - $off;
165            $lon_end = $lon + 5 + $off;
166
167            $pattern = "LDEM_{$LDEM}_lat_{$lat_s}_{$lat_e}_lon_{$lon_s}_{$lon_e}";
168            $special = FALSE;
169
170            if(file_exists("{$path_patternDB}/pattern_{$pattern}.inc"))
171            {
172                cline("Pattern {$pattern}: Pattern already exists - SKIPPING.");
173            }
174            else
175            {
176                cline("Pattern {$pattern}: Requesting data for pattern from database...");
177
178                if($lon_start < 0)
179                {
180                    $special = TRUE;
181                    $lon_extra = $lon_start + 360;
182                    $sql = "SELECT lat, lon, x, y, z
183                            FROM LDEM_{$LDEM}
184                            WHERE MBRContains(GeomFromText('POLYGON(({$lat_start} 0,
185                            {$lat_start} {$lon_end}, {$lat_end} {$lon_end}, {$lat_end} 0,
```

```php
                        {$lat_start} 0))'), point)
                    OR MBRContains(GeomFromText('POLYGON(({$lat_start}
                    {$lon_extra}, {$lat_start} 360, {$lat_end} 360, {$lat_end}
                    {$lon_extra}, {$lat_start} {$lon_extra}))'), point)
                ORDER BY point_id ASC;";
    }
    elseif($lon_end > 360)
    {
        $special = TRUE;
        $lon_extra = $lon_end - 360;
        $sql = "SELECT lat, lon, x, y, z
                FROM LDEM_{$LDEM}
                WHERE MBRContains(GeomFromText('POLYGON(({$lat_start}
                    {$lon_start}, {$lat_start} 360, {$lat_end} 360, {$lat_end}
                    {$lon_start}, {$lat_start} {$lon_start}))'), point)
                    OR MBRContains(GeomFromText('POLYGON(({$lat_start} 0,
                    {$lat_start} {$lon_extra}, {$lat_end} {$lon_extra},
                    {$lat_end} 0, {$lat_start} 0))'), point)
                ORDER BY point_id ASC;";
    }
    else
    {
        $sql = "SELECT lat, lon, x, y, z
                FROM LDEM_{$LDEM}
                WHERE MBRContains(GeomFromText('POLYGON(({$lat_start}
                    {$lon_start}, {$lat_start} {$lon_end}, {$lat_end} {$lon_end},
                    {$lat_end} {$lon_start}, {$lat_start} {$lon_start}))'), point)
                ORDER BY point_id ASC;";
    }
    
    $res = mysql_query($sql, $db_link) OR die(mysql_error($db_link));
    
    cline("Pattern {$pattern}: Creating data array...");
    $points = array();
    $csv = array();
    $point_id = 0;
    while($row = mysql_fetch_assoc($res))
    {
        $point_id++;
        $points[$point_id] = array("x" => $row["x"]/1000,
                                    "y" => $row["y"]/1000,
                                    "z" => $row["z"]/1000);
        $csv[] = "{$row["lat"]}, {$row["lon"]}\r\n";
    }
    unset($point_id);
    
    file_put_contents("{$path_tempdir}/pattern_{$pattern}.csv", $csv);
```

```
            cline("Pattern {$pattern}: Pattern (lat, lon) written to file.");
            unset($csv);

            if($lat_start < -90 OR $lat_end > 90)
            {
                $special = TRUE;
            }

            if($special)
            {
                cline("Pattern {$pattern}: Running Delaunay Triangulation...");
                exec("\"{$path_tempdir}/delaunay2D.exe\" pattern_{$pattern} 2>&1");
                cline("Pattern {$pattern}: Delaunay Triangulation finished.");

                $delaunay_template =
                    file("{$path_tempdir}/pattern_{$pattern}_delaunay.csv");
            }
            else
            {
                if(file_exists("{$path_tempdir}/delaunay_template_LDEM_{$LDEM}.csv"))
                {
                    $delaunay_template =
                        file("{$path_tempdir}/delaunay_template_LDEM_{$LDEM}.csv");
                    cline("Pattern {$pattern}: Delaunay Triangulation not necessary: ▼
                    Suitable template found.");
                }
                else
                {
                    cline("Pattern {$pattern}: No Delaunay template for standard case ▼
                    found. Running Delaunay Triangulation...");
                    exec("\"{$path_tempdir}/delaunay2D.exe\" pattern_{$pattern} 2>&1");
                    cline("Pattern {$pattern}: Delaunay Triangulation finished and ▼
                    template saved.");

                    rename("{$path_tempdir}/pattern_{$pattern}_delaunay.csv",
                    "{$path_tempdir}/delaunay_template_LDEM_{$LDEM}.csv");

                    $delaunay_template =
                        file("{$path_tempdir}/delaunay_template_LDEM_{$LDEM}.csv");
                }
            }

            cline("Pattern {$pattern}: Creating POV-Ray file using given Delaunay ▼
            Triangulation template...");
            $povrayfile = array(  "mesh\r\n",
                                  "{\r\n"     );
            for($i = 0; $i < count($delaunay_template); $i++)
```

```
                {
                    if(preg_match("/^([0-9]{1,}),([0-9]{1,}),([0-9]{1,})[\\s]{0,}$/",
                    $delaunay_template[$i], $matches))
                    {
                        $point_id_triang_1 = $matches[1];
                        $point_id_triang_2 = $matches[2];
                        $point_id_triang_3 = $matches[3];

                        $povrayfile[] = "    triangle\r\n";
                        $povrayfile[] = "    {\r\n";
                        $povrayfile[] = "        <{$points[$point_id_triang_1]["x"]}, ▼
                        {$points[$point_id_triang_1]["y"]}, ▼
                        {$points[$point_id_triang_1]["z"]}>, ▼
                        <{$points[$point_id_triang_2]["x"]}, ▼
                        {$points[$point_id_triang_2]["y"]}, ▼
                        {$points[$point_id_triang_2]["z"]}>, <▼
                        {$points[$point_id_triang_3]["x"]}, ▼
                        {$points[$point_id_triang_3]["y"]}, ▼
                        {$points[$point_id_triang_3]["z"]}>\r\n";
                        $povrayfile[] = "        texture { moon }\r\n";
                        $povrayfile[] = "    }\r\n";
                    }
                    else
                    {
                        die("Error: RegEx search in Delaunay template on line {$i} didn't ▼
                        succeeed.");
                    }
                }
                $povrayfile[] =          "}\r\n";

            file_put_contents("{$path_patternDB}/{$LDEM}/pattern_{$pattern}.inc",
            $povrayfile, LOCK_EX);
            unlink("{$path_tempdir}/pattern_{$pattern}.csv");
            if(file_exists("{$path_tempdir}/pattern_{$pattern}_delaunay.csv"))
            {
                unlink("{$path_tempdir}/pattern_{$pattern}_delaunay.csv");
            }
            cline("Pattern {$pattern}: Pattern written to POV-Ray file.");

            unset($povrayfile);
            unset($delaunay_template);
            unset($points);
        }
        $mtime = explode(" ",microtime());
        $mtime = $mtime[1] + $mtime[0];
        $endtime1 = $mtime;
        $totaltime1 = ($endtime1 - $starttime1);
```

```
            cline("Pattern {$pattern}: Calculation took {$totaltime1} sec.");
            cline();
        }
}

cline();
cline();
mysql_close($db_link) OR die(mysql_error($db_link));
cline("MySQL connection closed.");


    $mtime = microtime();
    $mtime = explode(" ",$mtime);
    $mtime = $mtime[1] + $mtime[0];
    $endtime = $mtime;
    $totaltime = ($endtime - $starttime);

cline();
cline("Operation has finished.");
cline("Peak memory needed: " . (memory_get_peak_usage()/(1024*1024)) . " MiB, total ▼
execution time: {$totaltime} sec.");

?>
```

Bibliography

Astrodynamics/Celestial Mechanics

[1] BRETAGNON, P.; FRANCOU, G.: *Planetary theories in rectangular and spherical variables – VSOP 87 solutions.* In: *Astronomy and Astrophysics* **202** (1-2): 309–315. IET, Stevenage, UK, 08/1988. ISSN 0004-6361. Online available at http://articles.adsabs.harvard.edu/cgi-bin/nph-iarticle_query?1988A%26A...202. .309B&data_type=PDF_HIGH&type=PRINTER&filetype=.pdf.

[2] EFROIMSKY, Michael: *Equations for the Keplerian Elements: Hidden Symmetry.* Preprint No. 1844 of the IMA (Institute of Applied Mathematics and its Applications), University of Minnesota, February 1, 2002. Online available at http://adsabs.harvard.edu/abs/2002ima..rept....1E.

[3] H.M. NAUTICAL ALMANAC Office (ed.): *Explanatory Supplement to the Astronomical Ephemeris and the American Ephemeris and Nautical Almanac.* Fourth impression (with amendents). Her Majesty's Stationery Office, London, 1977. ISBN 0-11-880578-9.

[4] MONTENBRUCK, Oliver; GILL, Eberhard: *Satellite Orbits: Models, Methods, Applications.* 1st edition 2000, corrected 3rd printing. Springer, Berlin, Heidelberg, 2005. ISBN 978-3540672807.

[5] MURRAY, Carl D.; DERMOTT, Stanley F.: *Solar System Dynamics.* Fourth printing, 2006. Cambridge University Press, New York, USA, 1999. ISBN 978-0-521-57597-3.

[6] NASA (ed.): *A Standardized Lunar Coordinate System for the Lunar Reconnaissance Orbiter and Lunar Datasets.* LRO Project and LGCWG White Paper, Version 5. NASA, October 1, 2008. Online available at http://lunar.gsfc.nasa.gov/library/LunCoordWhitePaper-10-08.pdf.

[7] SEIDELMANN, P. K.; ARCHINAL, B. A.; A'HEARN, M. F.; CRUIKSHANK, D. P.; HILTON, J. L.; KELLER, H. U.; OBERST, J.; SIMON, J. L.; STOOKE, P.; THOLEN, D. J.; THOMAS, P. C.: *Report Of The IAU/IAG Working Group On Cartographic Coordinates And Rotational Elements: 2003.* In: *Celestial Mechanics and Dynamical Astronomy* **91** (3-4): 203–215. Springer,

2005. ISSN 0923-2958. DOI 10.1007/s10569-004-3115-4. Preprint available online: http://astrogeology.usgs.gov/Projects/WGCCRE/WGCCRE2003preprint.pdf.

[8] SEIDELMANN, P. Kenneth (ed.): *Explanatory Supplement to the Astronomical Almanac*. First paperback impression. University Science Books, Sausalito, California, USA, 2006. ISBN 978-1-891389-45-0.

[9] SEIDELMANN, P. Kenneth; ARCHINAL, B. A.; A'HEARN, M. F.; CONRAD, A.; CONSOLMAGNO, G. J.; HESTROFFER, D.; HILTON, J. L.; KRASINSKY, G. A.; NEUMANN, G.; OBERST, J.; STOOKE, P.; TEDESCO, E. F.; THOLEN, D. J.; THOMAS, P. C.; WILLIAMS, I. P.: *Report of the IAU/IAG Working Group on cartographic coordinates and rotational elements: 2006*. In: *Celestial Mechanics and Dynamical Astronomy* **98** (3): 155–180. Springer, 2007. ISSN 0923-2958. DOI 10.1007/s10569-007-9072-y. Open Access: http://www.springerlink.com/content/e637756732j60270/fulltext.pdf.

[10] SIDI, Marcel J.: *Spacecraft Dynamics and Control: A Practical Engineering Approach*. First paperback edition, reprinted 2006. Cambridge University Press, New York, 2000. ISBN 978-0-521-78780-2.

[11] STANDISH, E. Myles; WILLIAMS, James G.: *Orbital Ephemerides of the Sun, Moon and Planets*. Online available at ftp://ssd.jpl.nasa.gov/pub/eph/planets/ioms/ExplSupplChap8.pdf. Retrieved 2011/05/15.

[12] WERTZ, James R. (ed.): *Spacecraft Attitude Determination and Control*. Astrophysics and Space Science Library **73**. First edition, tenth reprint, 2002. Kluwer Academic Publishers, Dordrecht, The Netherlands, 1978. ISBN 978-9027712042.

[13] WIKIPEDIA (ed.): *Argument of periapsis*. Online available at http://en.wikipedia.org/w/index.php?title=Argument_of_periapsis&oldid=450370870. Retrieved 2011/10/29.

[14] WIKIPEDIA (ed.): *Eccentric anomaly*. Online available at http://en.wikipedia.org/w/index.php?title=Eccentric_anomaly&oldid=416453169. Retrieved 2011/10/29.

[15] WIKIPEDIA (ed.): *Eccentricity vector*. Online available at http://en.wikipedia.org/w/index.php?title=Eccentricity_vector&oldid=443226923. Retrieved 2011/10/29.

[16] WIKIPEDIA (ed.): *Longitude of the ascending node*. Online available at http://en.wikipedia.org/w/index.php?title=Longitude_of_the_ascending_node&oldid=453399307. Retrieved 2011/10/29.

[17] WIKIPEDIA (ed.): *Orbital eccentricity*. Online available at http://en.wikipedia.org/w/index.php?title=Orbital_eccentricity&oldid=453223480. Retrieved 2011/10/29.

[18] WIKIPEDIA (ed.): *Orbital inclination*. Online available at http://en.wikipedia.org/w/index.php?title=Orbital_inclination&oldid=452407742. Retrieved 2011/10/29.

[19] WIKIPEDIA (ed.): *True anomaly*. Online available at http://en.wikipedia.org/w/index.php?title=True_anomaly&oldid=427251318. Retrieved 2011/05/15.

[20] WILLIAMS, J. G.; BOGGS, D. H.; FOLKNER, W. M.: *DE421 Lunar Orbit, Physical Librations, and Surface Coordinates*. JPL Interoffice Memorandum IOM 335-JW,DB,WF-20080314-001. JPL/California Institute of Technology, March 14, 2008. Online available at ftp://ssd.jpl.nasa.gov/pub/eph/planets/ioms/de421_moon_coord_iom.pdf.

Computer Vision (General)

[21] BREDIS, Kristian; LORENZ, Dirk: *Mathematische Bildverarbeitung – Einführung in Grundlagen und moderne Theorie*. 1. Auflage. Vieweg+Teubner Verlag, Wiesbaden, 2011. ISBN 978-3-8348-1037-3.

[22] GLASSNER, Andrew S. (ed.): *An Introduction to Ray Tracing*. Eighth printing. Morgan Kaufmann, San Francisco, California, USA, 2000. ISBN 978-0122861604.

[23] PHARR, Matt; HUMPHREYS, Greg: *Physically Based Rendering: From Theory to Implementation*. Second edition. Morgan Kaufmann, Elsevier, Burlington, MA, USA, 2010. ISBN 978-0-12-375079-2.

[24] SUFFERN, Kevin: *Ray Tracing from the Ground Up*. A K Peters, Wellesley, Massachusetts, USA, 2007. ISBN 978-1568812724.

Illuminance Flow Estimation

[25] BROOKS, Michael J.; HORN, Berthold K. P.: *Shape: And source from shading*. In: *Proceedings of the Ninth International Joint Conference on Artificial Intelligence* 2: 932–936. International Joint Conferences on Artificial Intelligence Organization, USA, 1985. Online available at http://ijcai.org/Past%20Proceedings/IJCAI-85-VOL2/PDF/052.pdf.

[26] CHANTLER, M.; PETROU, M.; PENIRSCHE, A.; SCHMIDT, M.; McGUNNIGLE, G.: *Classifying Surface Texture while Simultaneously Estimating Illumination Direction*. In: *International Journal of Computer Vision* **62** (1/2): 83–96. Springer Science, Netherlands, 2005. ISSN 0920-5691. DOI 10.1023/B:VISI.0000046590.98379.19.

[27] CHANTLER, M. J.: *Why illuminant direction is fundamental to texture analysis*. In: *IEE Proceedings of Vision, Image and Signal Processing* **142** (4): 199–206. 08/1995. ISSN 1350-245X. DOI 10.1049/ip-vis:19952065.

[28] CHANTLER, M. J.; DELGUSTE, G. B.: *Illuminant-tilt estimation from images of isotropic texture*. In: *IEE Proceedings of Vision, Image and Signal Processing* **144** (4): 213–219. 08/1997. ISSN 1350-245X. DOI 10.1049/ip-vis:19971302.

[29] CHANTLER, Mike; McGUNNIGLE, Ged; PENIRSCHKE, Andreas; PETROU, Maria: *Estimating Lighting Direction and Classifying Textures*. In: *Proceedings of the 2002 British Machine Vision Conference, Cardiff* : 737–746. The British Machine Vision Association and Society for Pattern Recognition, UK, 2001. Online available at http://www.bmva.org/bmvc/2002/papers/79/full_79.pdf.

[30] CHOJNACKI, Wojciech; BROOKS, Michael J.; GIBBINS, Danny: *Revisiting Pentland's estimator of light source direction*. In: *Journal of the Optical Society of America A* **11** (1): 118–124. The Optical Society, Washington, DC, USA, 1994. DOI 10.1364/JOSAA.11.000118.

[31] CHOW, Chi Kin; YUEN, Shiu Yin: *Illumination direction estimation for augmented reality using a surface input real valued output regression network*. In: *Pattern Recognition* **43** (4): 1700–1716. Pattern Recognition Society, Elsevier, April 2010. DOI 10.1016/j.patcog.2009.10.008.

[32] HARA, Kenji; NISHINO, Ko; IKEUCHI, Katsushi: *Light Source Position and Reflectance Estimation from a Single View without The Distant Illumination Assumption*. In: *IEEE Transactions on Pattern Analysis and Machine Intelligence* **27** (4): 493–505. IEEE, Los Alamitos, CA, USA, April 2005. ISSN 0162-8828. DOI 10.1109/TPAMI.2005.82.

[33] HORN, Berthold: *Obtaining shape from shading information*. In: WINSTON, P. H. (ed.): *The Psychology of Computer Vision*: 115–155. McGraw-Hill, New York, USA, 04/1975. ISBN 978-0070710481. Online available at http://people.csail.mit.edu/bkph/articles/Shape_from_Shading.pdf. Retrieved 2011/04/08.

[34] KARLSSON, Dan Stefan Mikael: *Illuminance Flow*. Thesis. Utrecht University, Netherlands, 2010. Online available at http://igitur-archive.library.uu.nl/dissertations/2010-0114-200210/karlsson.pdf.

[35] KARLSSON, Stefan; PONT, Sylvia; KOENDERINK, Jan: *Illuminance flow over anisotropic surfaces*. In: *Journal of the Optical Society of America A* **25** (2): 282–291. The Optical Society of America, Washington, DC, USA, February 2008. ISSN 1084-7529. DOI 10.1364/JOSAA.25.000282.

[36] KARLSSON, Stefan M.; PONT, Sylvia; KOENDERINK, Jan: *Illuminance flow over anisotropic surfaces with arbitrary viewpoint*. In: *Journal of the Optical Society of America A* **26** (5): 1250–1255. The Optical Society of America, Washington, DC, USA, May 2009. ISSN 1084-7529. DOI 10.1364/JOSAA.26.001250.

[37] KARLSSON, Stefan M.; PONT, Sylvia C.; KOENDERINK, Jan J.; ZISSERMAN, Andrew: *Illuminance Flow Estimation by Regression*. In: *International Journal of Computer Vision* **90** (3): 304–312. Springer, 12/2010. ISSN 0920-5691 (printed), 1573-1405 (electronic). DOI 10.1007/s11263-010-0353-7.

[38] KNILL, David C.: *Estimating illuminant direction and degree of surface relief*. In: *Journal of the Optical Society of America A* **7** (4): 759–775. The Optical Society, Washington, DC, USA, 04/1990. ISSN 1084-7529. DOI 10.1364/JOSAA.7.000759.

[39] KOENDERINK, Jan J.; PONT, Sylvia C.: *Irradiation direction from texture*. In: *Journal of the Optical Society of America A* **20** (10): 1875–1882. The Optical Society of America, Washington, DC, USA, October 2003. ISSN 1084-7529. DOI 10.1364/JOSAA.20.001875.

[40] KOENDERINK, Jan J.; VAN DOORN, Andrea J.; KAPPERS, Astrid M. L.; TE PAS, Susan F.; PONT, Sylvia C.: *Illumination direction from texture shading*. In: *Journal of the Optical Society of America A* **20** (6): 987–995. Optical Society of America, June 2003. ISSN 1084-7529. DOI 10.1364/JOSAA.20.000987.

[41] KOENDERINK, Jan J.; VAN DOORN, Andrea J.; PONT, Sylvia C.: *Perception of illuminance flow in the case of anisotropic rough surfaces*. In: *Attention, Perception, & Psychophysics* **69** (6): 895–903. Springer, New York, 2007. ISSN 1943-3921. DOI 10.3758/BF03193926.

[42] LI, Yuanzhen; LIN, Stephen; LU, Hanqing; SHUM, Heung-Yeung: *Multiple-cue Illumination Estimation in Textured Scenes*. In: *IEEE Proceedings of the 9th International Conference on Computer Vision* **2**: 1366–1373. IEEE, Nice, France, October 2003. ISBN 0-7695-1950-4. DOI 10.1109/ICCV.2003.1238649.

[43] LLADÓ, X.; OLIVER, A.; PETROU, M.; FREIXENET, J.; MARTÍ, J.: *Simultaneous surface texture classification and illumination tilt angle prediction*. Research paper for the 2003 British Machine Vision Conference, Norwich. The British Machine Vision Association and Society for Pattern Recognition, UK, 2003. Online available at http://www.bmva.org/bmvc/2003/papers/34/paper034.pdf.

[44] MAKI, Atsuto: *Estimation of illuminant direction and surface reconstruction by Geotensity constraint*. In: *Pattern Recognition Letters* **21** (13-14): 1115–1123. Elsevier Science B.V., Netherlands, 12/2000. ISSN 0167-8655. DOI 10.1016/S0167-8655(00)00072-6.

[45] NILLIUS, Peter; EKLUNDH, Jan-Olof: *Automatic Estimation of the Projected Light Source Direction*. In: *2001 IEEE Computer Society Conference on Computer Vision and Pattern Recognition (CVPR'01)* **1**: I-1076–I-1083. IEEE Computer Society, Los Alamitos, CA, USA, 2001. ISSN 1063-6919. ISBN 0-7695-1272-0. DOI 10.1109/CVPR.2001.990650.

[46] PENTLAND, Alex P.: *Finding the illuminant direction*. In: *Journal of the Optical Society of America* **72** (4): 448–455. The Optical Society of America, Washington, DC, USA, April 1982. DOI 10.1364/JOSA.72.000448.

[47] PONT, Sylvia; KOENDERINK, Jan: *Surface Illuminance Flow*. In: *Proceedings of the 2nd International Symposium on 3D Data Processing, Visualization, and Transmission (3DPVT'04)* : 2–9. IEEE Computer Society, Thessaloniki, Greece, 2004. DOI 10.1109/TDPVT.2004.1335134.

[48] PONT, Sylvia C.; KOENDERINK, Jan J.: *Illuminance Flow*. In: *Computer Analysis of Images and Patterns* : 90–97. Lecture Notes in Computer Science **2756**. Springer-Verlag, Berlin, Heidelberg, 2003. DOI 10.1007/978-3-540-45179-2_12.

[49] PONT, Sylvia C.; KOENDERINK, Jan J.: *Irradiation Orientation from Obliquely Viewed Texture*. In: *Deep Structure, Singularities, and Computer Vision* : 205–210. Lecture Notes in Computer Science **3753**. Springer-Verlag, Berlin, Heidelberg, 2005. DOI 10.1007/11577812_18.

[50] SAMARAS, Dimitrios; METAXAS, Dimitris: *Coupled Lighting Direction and Shape Estimation from Single Images*. In: *IEEE Proceedings of the 7th International Conference on Computer Vision* **2**: 868–874. IEEE, Kerkyra, Greece, September 1999. ISBN 0-7695-0164-8. DOI 10.1109/ICCV.1999.790313.

[51] STAUDER, Jürgen: *Point Light Source Estimation from Two Images and Its Limits*. In: *International Journal of Computer Vision* **36** (3): 195–220. Kluwer Academic Publishers, Boston, 2000. DOI 10.1023/A:1008177019313.

[52] VARMA, Manik; ZISSERMAN, Andrew: *Estimating Illumination Direction from Textured Images*. In: *2004 IEEE Computer Society Conference on Computer Vision and Pattern Recognition (CVPR'04)* **1**: 179–186. IEEE Computer Society, Los Alamitos, CA, USA, 2004. ISSN 1063-6919. DOI 10.1109/CVPR.2004.95.

[53] WEBER, Martin; CIPOLLA, Roberto: *A Practical Method for Estimation of Point Light-Sources*. In: *Proceedings of the 2001 British Machine Vision Conference, Manchester* **2**: 471–480. The British Machine Vision Association and Society for Pattern Recognition, UK, 2001. Online available at http://www.bmva.org/bmvc/2001/papers/117/accepted_117.pdf.

[54] WONG, Kwan-Yee; SCHNIEDERS, Dirk; LI, Shuda: *Recovering Light Directions and Camera Poses from a Single Sphere*. In: *Proceedings of the 10th European Conference on Computer Vision: Part I (ECCV'08), Marseille, France* : 631–642. Springer-Verlag, Berlin, Heidelberg, 2008. ISBN 978-3-540-88681-5. DOI 10.1007/978-3-540-88682-2_48.

[55] ZHENG, Qinfen; CHELLAPPA, Rama: *Estimation of Illuminant Direction, Albedo, and Shape from Shading*. In: *IEEE Transactions on Pattern Analysis and Machine Intelligence* **13** (7): 680–702. IEEE Computer Society, Los Alamitos, CA, USA, 1991. ISSN 0162-8828. DOI 10.1109/34.85658.

[56] ZHOU, Wei; KAMBHAMETTU, Chandra: *Estimation of Illuminant Direction and Intensity of Multiple Light Sources*. In: *Proceedings of the 7th European Conference on Computer Vision: Part IV (ECCV'02), Copenhagen, Denmark* : 206–220. Lecture Notes in Computer Science **2353**. Springer-Verlag, Berlin, Heidelberg, 2006. DOI 10.1007/3-540-47979-1_14.

Mathematics/Physics in General, Numerical Analysis and Computational Science

[57] ABRAMOWITZ, Milton; STEGUN, Irene A.: *Handbook of Mathematical Functions*. 10th printing. U.S. Department of Commerce, National Bureau of Standards, Washington, D.C., USA, December 1972

[58] ENGELN-MÜLLGES, Gisela; SCHÄFER, Wolfgang; TRIPPLER, Gisela: *Kompaktkurs Ingenieurmathematik. Mit Wahrscheinlichkeitsrechnung und Statistik*. 3., neu bearbeitete und erweiterte Auflage. Fachbuchverlag Leipzig, 2004. ISBN 9783446228641.

[59] HANKE-BOURGEOIS, Martin: *Grundlagen der Numerischen Mathematik und des Wissenschaftlichen Rechnens*. 3., aktualisierte Auflage. Vieweg+Teubner Verlag, Wiesbaden, 2009. ISBN 978-3-8348-0708-3.

[60] HILBERT, Alfred: *Mathematik*. 1. Auflage. VEB Fachbuchverlag Leipzig, 1987. ISBN 3-343-00248-8.

[61] JAWORSKI, Boris M.; DETLAF, A. A.: *Physik-Handbuch für Studium und Beruf.* Verlag Harri Deutsch, Thun/Frankfurt/Main, 1986. ISBN 3-87-144-804-4.

[62] KÖRNER, Wolfgang; HAUSMANN, Ewald; KIEßLING, Günther; MENDE, Dietmar; SPRETKE, Hellmut: *Physik: Fundament der Technik*. 10. Auflage. VEB Fachbuchverlag Leipzig, 1989. ISBN 3-343-00240-2.

[63] KUIPERS, Jack B.: *Quaternions and Rotation Sequences: A Primer with Applications to Orbits, Aerospace, and Virtual Reality*. 5th printing, and first paperback printing. Princeton University Press, Princeton, New Jersey, USA, 2002. ISBN 978-0-691-10298-6.

[64] MERZINGER, Gerhard; MÜHLBACH, Günter; WILLE, Detlef; WIRTH, Thomas: *Formeln und Hilfen zur höheren Mathematik*. 4. Auflage. Binomi Verlag, Springe, Oktober 2004. ISBN 3-923923-35-X.

[65] MESCHEDE, Dieter: *Gerthsen Physik*. 23., überarbeitete Auflage. Springer, Berlin, 2006. ISBN 978-3-540-25421-8.

[66] PAPULA, Lothar: *Mathematik für Ingenieure und Naturwissenschaftler. Band 1*. 10., erweiterte Auflage. Vieweg, Braunschweig/Wiesbaden, Oktober 2001. ISBN 3-528-94236-3.

[67] WIKIPEDIA (ed.): *Quaternions and spatial rotation*. Online available at http://en.wikipedia.org/w/index.php?title=Quaternions_and_spatial_rotation&oldid=484703420. Retrieved 2012/04/02.

[68] ZURMÜHL, Rudolf: *Matrizen und ihre technischen Anwendungen*. 4., neubearbeitete Auflage. Springer, Berlin, 1964

Scientific Tables/Conventions, Works of Reference, Algorithms

[69] LEY, Wilfried (ed.); WITTMANN, Klaus (ed.); HALLMANN, Willi (ed.): *Handbuch der Raumfahrttechnik*. 3., völlig neu bearbeitete und erweiterte Auflage. Hanser Verlag, München, 2008. ISBN 9783446411852.

[70] MEEUS, Jean: *Astronomical Algorithms*. Second edition, 1998. Willmann-Bell, Richmond, Virginia, USA, 2009. ISBN 9780943396613.

[71] MÜLLER, Edith A. (ed.): *Proceedings of the Sixteenth General Assembly, Grenoble 1976*. Transactions of the International Astronomical Union **XVIB**. International Astronomical Union, Kluwer Academic Publishers, 1977. ISBN 90-277-0836-3.

[72] MOHR, Peter J.; TAYLOR, Barry N.: *CODATA recommended values of the fundamental physical constants: 1998*. In: *Review of Modern Physics* **72** (2): 351–495. American Physical Society, April 2000. DOI 10.1103/RevModPhys.72.351.

[73] MOHR, Peter J.; TAYLOR, Barry N.: *CODATA recommended values of the fundamental physical constants: 2002*. In: *Review of Modern Physics* **77** (1): 1–107. American Physical Society, January 2005. DOI 10.1103/RevModPhys.77.1.

[74] MOHR, Peter J.; TAYLOR, Barry N.; NEWELL, David B.: *CODATA recommended values of the fundamental physical constants: 2006*. In: *Review of Modern Physics* **80** (2): 633–730. American Physical Society, 2008. ISSN 1539-0756. DOI 10.1103/RevModPhys.80.633. Online available at http://physics.nist.gov/cuu/Constants/RevModPhys_80_000633acc.pdf.

[75] NASA/JPL (ed.): *Astrodynamic Constants*. NASA/JPL/SSD website, 2011. Online available at http://ssd.jpl.nasa.gov/?constants. Retrieved 2011/04/26.

[76] NASA/JPL (ed.): *Planetary Data System Standards Reference*. Version 3.8. Jet Propulsion Laboratory, JPL D-7669, Part 2. JPL/California Institute of Technology, February 27, 2009. Online available at http://pds.nasa.gov/tools/standards-reference.shtml. Retrieved 2012/03/19.

[77] U.S. NAUTICAL ALMANAC Office (ed.): *The Astronomical Almanac for the Year 2011*. U.S. Government Printing Office, Washington, 2010. ISBN 978-0-7077-41031.

[78] WIECZOREK, Mark A.; JOLLIFF, Bradley L.; KHAN, Amir; PRITCHARD, Matthew E.; WEISS, Benjamin P.; WILLIAMS, James G.; HOOD, Lon L.; RIGHTER, Kevin; NEAL, Clive R.; SHEARER, Charles K.; MCCALLUM, I. Stewart; TOMPKINS, Stephanie; HAWKE, B. Ray; PETERSON, Chris; GILLIS, Jeffrey J.; BUSSEY, Ben: *The Constitution and Structure of the Lunar Interior*. In: *Reviews in Mineralogy and Geochemistry* **60** (1): 221–364. 2006. DOI 10.2138/rmg.2006.60.3. Online available at http://scripts.mit.edu/~paleomag/articles/60_03_Wieczorek_etal.pdf.

Spacecraft Engineering

[79] FORTESCUE, Peter (ed.); STARK, John (ed.); SWINERD, Graham (ed.): *Spacecraft Systems Engineering*. Third edition. Wiley, England, 2003. ISBN 0-471-61951-5.

[80] WERTZ, James R. (ed.); LARSON, Wiley J. (ed.): *Space Mission Analysis and Design*. Space Technology Series **8**. Third edition, tenth printing, 2008. Microcosm Press & Springer, New York, 1999. ISBN 9780792359012.

Other Topics

[81] ARAKI, H.; TAZAWA, S.; NODA, H.; ISHIHARA, Y.; GOOSSENS, S.; SASAKI, S.; KAWANO, N.; KAMIYA, I.; OTAKE, H.; OBERST, J.; SHUM, C.: *Lunar Global Shape and Polar Topography Derived from Kaguya-LALT Laser Altimetry*. In: *Science* **323** (5916): 897–900. 2009. DOI 10.1126/science.1164146.

[82] BUNDESMINISTERIUM FÜR WIRTSCHAFT UND TECHNOLOGIE (ed.): *Für eine zukunftsfähige deutsche Raumfahrt: Die Raumfahrtstrategie der Bundesregierung*. Bundesministerium für Wirtschaft und Technologie, November 2010. Online available at http://www.bmwi.de/BMWi/Redaktion/PDF/B/zukunftsfaehige-deutsche-raumfahrt, property=pdf,bereich=bmwi,sprache=de,rwb=true.pdf.

[83] CHIN, Gordon; BRYLOW, Scott; FOOTE, Marc; GARVIN, James; KASPER, Justin; KELLER, John; LITVAK, Maxim; MITROFANOV, Igor; PAIGE, David; RANEY, Keith; ROBINSON, Mark; SANIN, Anton; SMITH, David; SPENCE, Harlan; SPUDIS, Paul; STERN, S.; ZUBER, Maria: *Lunar Reconnaissance Orbiter Overview: The Instrument Suite and Mission*. In: *Space Science Reviews* **129**: 391–419. Springer Netherlands, April, 2007. ISSN 0038-6308. DOI 10.1007/s11214-007-9153-y. Online available at http://lro.gsfc.nasa.gov/library/LRO_Space_Science_Paper.pdf.

[84] CONNOLLY, John F.: *Constellation Program Overview*. NASA, Constellation Program Office, October 2006. Online available at http://www.nasa.gov/pdf/163092main_constellation_program_overview.pdf.

[85] DLR (ed.): *Flug über den dreidimensionalen Mond*. Website of the German Aerospace Center (DLR), November 21, 2011. Online available at http://www.dlr.de/dlr/desktopdefault.aspx/tabid-10081/151_read-2065/. Retrieved 2012/04/01.

[86] ESA (ed.): *The Aurora Programme*. European Space Agency Publications Division, The Netherlands, 2004. Online available at http://esamultimedia.esa.int/docs/Aurora/Aurora625_2.pdf.

[87] ISECG (ed.): *The Global Exploration Roadmap*. International Space Exploration Coordination Group (ISECG), NASA, Washington, September 2011. Online available at http://www.nasa.gov/pdf/591067main_GER_2011_small_single.pdf.

[88] JOHNSON, Andrew E.; ANSAR, Adnan; MATTHIES, Larry H.; TRAWNY, Nikolas; MOURIKIS, Anastasios I.; ROUMELIOTIS, Stergios I.: *A General Approach to Terrain Relative Navigation for Planetary Landing*. In: *2007 AIAA Conference and Exhibit, May 7–10, 2007, Rohnert Park, California*: 1–9. American Institute of Aeronautics and Astronautics, 2007. Online available at http://www-users.cs.umn.edu/~stergios/papers/AIAA_Infotech07.pdf.

[89] JOHNSON, Andrew E.; MONTGOMERY, James F.: *Overview of Terrain Relative Navigation Approaches for Precise Lunar Landing*. In: *2008 IEEE Aerospace Conference, Big Sky, Montana, March 1, 2008*: 1–10. IEEEAC paper #1657, Version 2, Updated December 14, 2008.

Online available at http://trs-new.jpl.nasa.gov/dspace/bitstream/2014/41154/1/07-4467.pdf.

[90] MAASS, Bolko; KRÜGER, Hans; THEIL, Stephan: *An Edge-Free, Scale-, Pose- and Illumination-Invariant Approach to Crater Detection for Spacecraft Navigation.* In: *Proceedings of the 7th International Symposium on Image and Signal Processing and Analysis (ISPA 2011), Dubrovnik, Croatia, September 4-6, 2011* : 603–608. University of Zagreb, Faculty of Electrical Engineering and Computing, Croatia, 2011. ISSN 1845-5921. ISBN 978-1-4577-0841-1.

[91] MORITZ, H.: *Geodetic Reference System 1980.* In: *Journal of Geodesy* **74**: 128–162. Springer, Berlin/Heidelberg, 2000. ISSN 0949-7714. DOI 10.1007/s001900050278. Online available at http://www.springerlink.com/content/0bgccvjj5bedgdfu/fulltext.pdf.

[92] MYSQL (ed.): *MySQL 5.5 Reference Manual — 11.17.6.1. Creating Spatial Indexes.* Online available at http://dev.mysql.com/doc/refman/5.5/en/creating-spatial-indexes.html. Retrieved March 19, 2012.

[93] MYSQL (ed.): *MySQL 5.5 Reference Manual — C.5.2.10 Packet too large.* Online available at http://dev.mysql.com/doc/refman/5.5/en/packet-too-large.html. Retrieved March 19, 2012.

[94] MYSQL (ed.): *MySQL 5.5 Reference Manual — Chapter 13: Storage Engines.* Online available at http://dev.mysql.com/doc/refman/5.5/en/storage-engines.html. Retrieved January 15, 2012.

[95] MYSQL (ed.): *MySQL 5.5 Reference Manual — Section 11.17: Spatial Extensions.* Online available at http://dev.mysql.com/doc/refman/5.5/en/spatial-extensions.html. Retrieved January 15, 2012.

[96] NASA (ed.): *The Vision for Space Exploration.* NASA, February 2004. Online available at http://www.nasa.gov/pdf/55583main_vision_space_exploration2.pdf.

[97] NASA (ed.): *2011 NASA Strategic Plan.* NASA, 2011. Online available at http://www.nasa.gov/pdf/516579main_NASA2011StrategicPlan.pdf.

[98] NASA NAIF (ed.): *NAIF CSPICE Toolkit Hypertext Documentation: spkpos_c.* NASA, The Navigation and Ancillary Information Facility (NAIF), June 9, 2010. Online available at http://naif.jpl.nasa.gov/pub/naif/toolkit_docs/C/cspice/spkpos_c.html. Retrieved 2012/03/31.

Bibliography

[99] NASA NAIF (ed.): *SPICE Tutorials (Merged)*. NASA, The Navigation and Ancillary Information Facility (NAIF), January 2012. Online available at ftp://naif.jpl.nasa.gov/pub/naif/toolkit_docs/Tutorials/pdf/packages/SPICE_Tutorials_Merged.pdf. Retrieved 2012/03/31.

[100] NASA Office of Public Affairs (ed.): *Global Exploration Strategy and Lunar Architecture*. NASA, Johnson Space Center, December 4, 2006. Online available at http://www.nasa.gov/pdf/164021main_lunar_architecture.pdf.

[101] NASA (ed.): *The Moon's hourly appearance in 2012*. Website of the NASA Lunar Science Institute, March 5, 2012. Online available at http://lunarscience.nasa.gov/articles/the-moons-hourly-appearance-in-2012/. Retrieved 2012/04/01.

[102] Neumann, Gregory A.: *Lunar Reconnaissance Orbiter Lunar Orbiter Laser Altimeter: Reduced Data Record and Derived Products — Software Interface Specification*. Version 2.2. NASA Goddard Space Flight Center, Greenbelt, April 28, 2009. Online available at http://lunar.gsfc.nasa.gov/lola/images/LOLA_RDRSIS.pdf. Retrieved 2012/03/19.

[103] POV-Ray (ed.): *Newsgroup povray.windows: TRUE batch mode*. Newsgroup thread at povray.windows, 2011. Online available at http://news.povray.org/povray.windows/thread/%3Cweb.4d9c0910b00cf4ef6c1e98510@news.povray.org%3E/. Retrieved 2012/04/01.

[104] POV-Ray (ed.): *POV-Ray 3.6 Documentation Online View: 1.4.4.7 Why are triangle meshes in ASCII format?* POV-Ray website, 2008. Online available at http://www.povray.org/documentation/view/3.6.1/171/. Retrieved 2012/04/01.

[105] POV-Ray (ed.): *POV-Ray Reference for POV-Ray Version 3.6.1*. POV-Ray website, August 2004. Online available at http://www.povray.org/download/. Retrieved 2012/04/01.

[106] Schrunk, David G.; Sharpe, Burton L.; Cooper, Bonnie L.; Thangavelu, Madhu: *The Moon: Resources, Future Development, and Settlement*. Second Edition. Springer/Praxis Publishing, Chichester, UK, 2008. ISBN 978-0-387-36055-3.

[107] Smith, David; Zuber, Maria; Jackson, Glenn; Cavanaugh, John; Neumann, Gregory; Riris, Haris; Sun, Xiaoli; Zellar, Ronald; Coltharp, Craig; Connelly, Joseph; Katz, Richard; Kleyner, Igor; Liiva, Peter; Matuszeski, Adam; Mazarico, Erwan; McGarry, Jan; Novo-Gradac, Anne-Marie; Ott, Melanie; Peters, Carlton; Ramos-Izquierdo, Luis; Ramsey, Lawrence; Rowlands, David; Schmidt, Stephen; Scott, V.; Shaw, George; Smith, James; Swinski, Joseph-Paul; Torrence, Mark; Unger, Glenn; Yu, Anthony; Zagwodzki, Thomas: *The Lunar Orbiter Laser Altimeter Investigation*

on the Lunar Reconnaissance Orbiter Mission. In: *Space Science Reviews* **150**: 209–241. Springer Netherlands, 2010. ISSN 0038-6308. DOI 10.1007/s11214-009-9512-y.

[108] WIKIPEDIA (ed.): *Boris Delaunay.* Online available at http://en.wikipedia.org/w/index.php?title=Boris_Delaunay&oldid=473408519. Retrieved 2012/04/01.

[109] WIKIPEDIA (ed.): *Delaunay-Triangulation.* Online available at http://de.wikipedia.org/w/index.php?title=Delaunay-Triangulation&oldid=95943619. Retrieved 2012/04/01.

[110] WIKIPEDIA (ed.): *File:Delaunay circumcircles centers.png.* Online available at http://en.wikipedia.org/wiki/File:Delaunay_circumcircles_centers.png. Retrieved 2012/04/01.

[111] WIKIPEDIA (ed.): *File:Dolphin triangle mesh.svg.* Online available at http://commons.wikimedia.org/wiki/File:Dolphin_triangle_mesh.svg. Retrieved 2012/04/01.

[112] WIKIPEDIA (ed.): *File:Mesh overview.svg.* Online available at http://en.wikipedia.org/wiki/File:Mesh_overview.svg. Retrieved 2012/04/01.

[113] WIKIPEDIA (ed.): *William Rowan Hamilton.* Online available at http://en.wikipedia.org/w/index.php?title=William_Rowan_Hamilton&oldid=483319860. Retrieved 2012/04/02.

[114] WIKTIONARY (ed.): *verto.* Online available at http://en.wiktionary.org/w/index.php?title=verto&oldid=16356726. Retrieved 2012/04/01.

Listings

List of Figures

1.1 TRON facility in an early and advanced stage of construction. © DLR; pictures reproduced with friendly permission. 19

1.2 A sample 3D surface tile of the Moon used in an advanced construction stage of TRON, illuminated with the 5 DOF illumination system: Realistic shadows are cast. © DLR; picture reproduced with friendly permission. 20

1.3 Schematic overview of the *Testbed for Robotic Optical Navigation* (TRON) laboratory, located at the DLR Institute of Space Systems in Bremen, as seen from above. The lab is divided into two sections: Operations (the simulation control room) and simulations section. At the bottom and the right border of the simulation section's illustration the wall-mounted surfaces tiles are depicted. These tiles can be illuminated by a 5 DOF illumination system, as implied by the red beam. The scene is then captured with sensors, i.e. optical or LIDAR sensors, which are mounted at the tool center point of the 6 DOF industrial robot on a rail system. Reproduced with friendly permission from [90]. 21

1.4 Extraction of crater contrast areas out of a sample picture; the vector shown beyond is the local solar illumination vector. © DLR; reproduced from [90, p. 605] with friendly permission. 22

1.5 First results of the new crater detection algorithm on pictures of real celestial surfaces. © DLR; pictures reproduced from [90, p. 608] with friendly permission. 23

1.6 Overview of active-sensing approaches for TRN. Only shown in completion to figure 1.7, since these methods rely on ranging techniques, which are not applicable to the thesis software. (Image based on the table in [89, p. 4]) 25

1.7 Overview of passive-sensing approaches for terrain-relative navigation (TRN). The thesis software can produce renderings of planetary surfaces for all kinds of passive-sensing approaches for TRN, indicated by fields with yellow and orange backgrounds. The crater detection algorithm currently developed by DLR within the ATON/TRON project uses the Crater Pattern Matching approach, which is marked in orange. As it can be inferred from this overview, the thesis software can be used not only for the current DLR project, but for all other TRN approaches. (Image based on the table in [89, p. 4]) 26

2.1 Overview of the components of the *Moon Surface Illumination Simulation Framework* (MSISF). 31

2.2 Visualization of the difference between a digital terrain model (DTM) and a digital surface model (DSM). 32

3.1 Two distinct points on the surface of a sphere with center in the coordinate origin $\mathbf{O} = (0,0,0)^{\mathrm{T}}$ and radius r. 41

3.2 Representation of an arbitrary point $\mathbf{p}$ using spherical coordinates with polar angle α, azimuthal angle φ and the radial distance r to the orgin of the coordinate system. 41

4.1 Plane visualization of the Clementine topographic data at the finest resolution (4 px/deg). A visual examination of the data shows that this data set is impractical to generate a realistic surface illumination. 46

4.2 The Lunar Reconnaissance Orbiter (LRO) in a near-final construction stage. The entire instrument suite is visible from this perspective; the Lunar Orbiter Laser Altimeter (LOLA) is the conical instrument directly beyond the white shining plate. © NASA/Debbie McCallum. Obtained from http://www.nasa.gov/mission_pages/LRO/multimedia/lrocraft5.html. 48

4.3 LOLA laser geometry on the ground for four consecutive shots; the numbers represent the channel numbers. The solid, black-filled circles indicate the transmitted laser footprints on the Moon's surface, while the solid, concentric circles imply the receiver's field of view. © NASA. Obtained from http://lunar.gsfc.nasa.gov/lola/images/fig.pdf. 49

4.4 Visible flaws in the surface data caused by data interpolation in the preliminary LRO LOLA LDEM products, shown on an example MSIS rendering. 51

5.1 Overview of mesh elements in computer graphics. Source: [112], License: Creative Commons Attribution-Share Alike 3.0 Unported. 68

List of Figures

5.2 Result of a 2D DELAUNAY triangulation for an example point set. The vertices are the black filled dots, and the edges are black lines. The gray circles represent the circumferences of each resulting triangle, while the red filled dots are the center points of each circumference. Vectorized version of [110] by Matthias Kopsch; License: Creative Commons Attribution-Share Alike 3.0 Unported. . . . 69

5.3 A triangulated dolphin. Source: [111]; License: Public Domain. 70

8.1 Example ray tracing scene with a sphere. The sphere's center is placed at the origin $(0, 0, 0)^T$ of the right-handed coordinate system. Three example rays s_1, s_2 and s_3 are shot: s_1 misses the sphere, s_2 is tangential to the sphere's surface and s_3 intersects the sphere twice. In ray 2 and 3, the intersection points can be calculated solving equation 8.4 for λ or λ_1 and λ_2, respectively, and substituting the parameter(s) into formula 8.2. 104

8.2 POV-Ray camera geometry to be used. The focal point c_{pos} of the camera is identical with the spacecraft/camera location; c_{pos} is a spatial coordinate within the ME/PA reference frame. The user-given field of view angle α ordains the length of the camera direction vector $c_{direction}$ and, by implication, the distance between the focal point c_{pos} and the image plane Ω.

The pixels of the later rendered picture are a strict subset of all points on the image plane Ω, which is defined by the camera-up vector c_{up} and the camera-right vector c_{right}. By definition, the rendered image is an axes-parallel rectangle with a width of $\|c_{right}\|$ and a height of $\|c_{up}\|$, while the axes, in this case, are c_{right} and c_{up} itself; the rectangle is centered at $c_{pos} + c_{direction}$.

c_{right} and c_{up} are influenced by the user-specified pixel width and height of the image to be rendered. The determination of the camera orientation is explained in chapter 7. 106

8.3 Illustration of pattern squeezing at the poles (the orange marked area): A rendering of the same surface area near the poles needs more surface patterns. The illustration is not drawn to scale (the MSISF surface patterns are much smaller). 113

8.4 DSPSA drawbacks visible on successive renderings of a series. The time is frozen (the Sun's position in relation to the Moon's position will not change), but the camera moves along a trajectory around the Moon. The camera moves to the image north with each picture (from left to right, top to bottom).

Each left picture shows a shadow caused by an object, which is not visible in the scene. Each right picture shows the next rendering, when the camera has moved a little bit upwards — some shadows disappear, because the patterns, on which the objects causing the shadows are located, are not selected anymore. The object, never visible itself, has disappeared, and so has its shadow. 114

9.1   MSIS rendering with rendering annotations activated. The red lines indicate the local solar illumination angle for each grid sample point, which is marked as a red dot at the beginning of the lines. The MSIS will only visualize the illumination angle on pixels, which show the Moon's surface (not the space background). 117
9.2   Schematic of the local illumination angle determination principle. 124
9.3   Geometrical construction of $\mathbf{p}_{local}$. 128
9.4   Geometrical construction of the local illumination angle α. 133

10.1 Current construction progress of the TRON facility. Visible is the floor-mounted 6 DOF industrial robot simulating a spacecraft as well as the ceiling-mounted 5 DOF lighting system. © DLR; picture reproduced with friendly permission. 141
10.2 A TRON 3D surface tile of the Moon's surface for a project on behalf of the European Space Agency (ESA). © DLR; picture reproduced with friendly permission. 142

List of Tables

1.1   This table shows common future targets of space exploration missions including their key objectives and challenges. ("Summary of the Destination Assessment Activity", quoted from [87, p. 15]) 16
2.1   Hardware configuration for the machine used as test and development environment. 35
4.1   Available LRO LOLA equi-rectangular map-projected LDEM products. Based on [107, p. 239]. 50
4.2   Data coverage of the LOLA LDEM products for versions 1.05 and 1.07. Source: Errata file of the LOLA PDS data node at http://imbrium.mit.edu/ERRATA.TXT. 52
4.3   MySQL query performance comparison for a 5° × 5° surface patch using standard indices (PRIMARY KEY(...)). 55
4.4   MySQL profiling for the slow query on the LDEM_64 table (all values in seconds). 55
4.5   Comparison of the storage requirements for the used LDEM resolutions distinguished by the intermediate and the final table layout. 57
4.6   MySQL query performance comparison for a 5° × 5° surface patch using a spatial index (SPATIAL KEY ...). 58

4.7	MySQL profiling for the optimized query on the LDEM_64 table (all values in seconds).	58
6.1	Overview of the required command-line arguments for the different operation modes of the MSIS. All arguments marked with ✔ are mandatory for the respective operation mode, while all arguments marked with ✗ are not allowed and those with ○ are optional. All additional arguments are optional and have been documented in appendix A.	93

List of Code Listings

4.1	Final table layout.	56
4.2	Example content of a PDS label file. Shown here: The corresponding label file `LDEM_1024_00N_15N_330_360.LBL` for a LDEM file (`.img` file), shortened only to give an impression of the PDS format.	63
6.1	Implementation of the Sun position calculation using CSPICE in C++.	82
6.2	Implementation of the spacecraft position calculation using KEPLERian elements into the MSIS.	87
6.3	Implementation of the conversion from state vectors to KEPLERian elements into the MSIS.	90
6.4	MSIS implementation of the modified Julian date to Gregorian date/time algorithm, based on Jean MEEUS' algorithm [70, p. 63].	91
8.1	MSIS code implementation of the Dynamical Surface Pattern Selection Algorithm (DSPSA).	111
8.2	DPSPA invocation within the `Simulation` class. This code snippet has to be executed for each image to be rendered.	112

Alphabetical Index

— Symbols —

i, 98
j, 98
k, 98

— A —

ATON project, 17 f.
    principal objective, 18
Augustine Commission, 14
Aurora Program, 16
autonomous navigation and landing, 16 f.
Autonomous Terrain-based Optical Navigation, *see* ATON project

— C —

C#, 30
camera
    camera right vector, 101
    initial orientation, 101
    orientation, 100
Clementine mission, 45 f.
complex numbers, 98
Constellation Program, 13 f.
crater(s), 20
    detection algorithm(s), 22
    detection treshold(s), 23
    imaging and detection, 22
CxP, *see* Constellation Program

— D —

Deep Space Program Science Experiment, *see* Clementine mission
DELAUNAY triangulation, 32, 68
DEM, *see* Digital Elevation Model(s)
Descent Orbit Injection maneuver, 17
Digital Elevation Model(s), 18, 23, 30, 45 f., 48
DLR, *see* German Aerospace Center
DOI, *see* Descent Orbit Injection maneuver
doxygen, 81
DSPSA, 33, 80, 103 f.
    implementation, 109–113
DSPSE, *see* Clementine mission
Dynamical Surface Pattern Selection Algorithm, *see* DSPSA

— E —

ESA, *see* European Space Agency
EULER angles, 97
European Space Agency, 16

— F —

flexible-path approach, 14
flight path control, 16 f.

— G —

GDR, 49

Alphabetical Index

German Aerospace Center, 17
    Institute of Space Systems, 17
German national space program, 16
gimbal lock, 97
Gregorian date, 90
Gridded Data Records, *see* GDR

— H —

HA, *see* Hazard Avoidance
Hamilton, Sir William Rowan, 98
Hazard Avoidance, 18

— I —

IDE, 79
in-situ
    position determination, 17
    utilization of resources, 16
integrated development environment, *see* IDE
International Space Station, 13
ISS, *see* International Space Station

— J —

JAXA SELENE mission, 46 f.
JD, 90
Julian date, *see* JD

— K —

Kepler's equation, 84–89
Keplerian elements, 84–89

— L —

landing place selection, 18

landmark based navigation, 17
LDEM, 32, 47, 49 f.
LEO, *see* Low Earth Orbit
LIDAR, 29
Light Detection and Ranging, *see* LIDAR
local solar illumination angle, 122–133
LOLA, 47 f., 50
    data products, 48, 51
    MySQL import process, 59–65
Low Earth Orbit, 13
LRO, 32, 47
    instrument suite (figure), 48
    laser geometry (figure), 49
Lunar Digital Elevation Model, *see* LDEM
lunar landing procedure, 17–20
Lunar Orbiter Laser Altimeter, *see* LOLA
Lunar Reconnaissance Orbiter, *see* LRO

— M —

Mars, 13–16
Mars-first approach, 13
MBR, *see* Minimum Bounding Rectangle
ME/PA, *see* Mean Earth/Polar Axis
Mean Earth/Polar Axis, 30, 39 f., 68, 97
mesh modeling
    edges, 67
    faces, 67
    surfaces, 67
    vertices, 67
Minimum Bounding Rectangle, 56
Mir, 13
MJD, 80, 90
modified Julian date, *see* MJD
Moon
    flattening, 40
    generalization as a sphere, 40
    selenograhic latitude, 43
    selenographic longitude, 43

Moon Surface Illumination Simulation
    Framework, *see* MSISF
Moon Surface Illumination Simulator, *see* MSIS
Moon-first approach, 13 f.
MSIS, 32, 50
    basic output, 115
    classes, 79 ff.
    date/time conversion algorithm, 90
    definiton of inputs/outputs, 36
    example usage, 94 f.
    invocation, 92 ff.
    `KeplerOrbit` class, 81
    `Program` class, 79
    rendering example, 51
    `Simulation` class, 79 f.
    software architecture, 79 ff.
    `Spacecraft` class, 80
    spacecraft position calculation, 84 ff.
    `SpacecraftState` class, 80
    state vector conversion, 87 ff.
    Sun position calculation, 81–84
    `tools` class, 80
    user interface, 36, 92 ff.
    XML meta information file, 116
    XML pre-defined tags/attributes, 119–122
MSISF, 51 f., 103
    components, 30 f.
    deployment/installation, 38
    development milestones, 30–33
    development/test environment, 34 f.
    file system layout, 37 f.
    general concept, 29 f.
    pattern repository, 30
    used software/programming languages, 34 f.
My Indexed Sequential Access Method, *see*
    MyISAM
My Structured Query Language, *see* MySQL
MyISAM, 53 f., 56
MySQL, 51–65
    server configuration, 57
    spatial extensions, 56

— **N** —

navigation technologies, 16 f.
near-Earth asteroids, 14 ff.

— **O** —

optical landing techniques, 17
optical navigation, 17, 24
orbit determination, 16
orbital elements
    Earth's orbital elements, 16
    spacecraft's orbital elements, 16

— **P** —

pattern generation, 71–74
    CSV file, 73 f.
    script, 72 ff.
    triangulation function, 74
PDI, *see* Power Descent Initiate
PDS, 49 f.
    required parameters, 49
Persistence of Vision Raytracer, *see* POV-Ray
Planetary Data System, *see* PDS
    required parameters, 60
point cloud, 67
POV-Ray, 30, 32, 103
    camera geometry, 105–108
    file splitting, 71
    mesh compilation, 74
    mesh support, 68
    scene description language (SDL), 70 f.
    script, 74, 76
    syntax (code snippet), 70 f.
Power Descent Initiate, 18

— **Q** —

quaternions, 97–101

3D vector representation, 99
conjugate, 99
field axioms, 99
multiplication, 99
norm, 99
notation, 98
pure quaternions, 98
rotation quaternions, 99
unit quaternion, 99

— R —

R-tree, 56
ray tracing, 104 f.
real-time simulation, 20
reference frame(s), 16, 30, 39 f., 54, 97
rendering annotation, 80
rotation matrices, 97

— S —

selenographic coordinate system, 31
Selenological and Engineering Explorer, see
    JAXA SELENE mission
Space Age, 13
space exploration, 13–17
    international goals, 14 f.
    key objectives and challenges, 15 f.
spacecraft
    orientation, 100 f.
    orientation model, 97–101
    rotation, 97–101
spatial rotation, 97–100
spherical coordinate system, 40–43
spherical polar coordinates, 42
SPICE, 81–84, 90
Sputnik 1, 13
state vectors, 84–89
surface illumination simulation, 24–27

surface pattern, 71
    anatomy, 67–71
    storage, 74

— T —

terrain-relative navigation, 18, 24
    active sensing approaches (figure), 25
    active/passive sensing, 24–27
    passive sensing approaches (figure), 26
Testbed for Robotic Optical Navigation, see
    TRON
thesis' objectives, 24–27
triangulated irregular networks, 32
TRN, see terrain-relative navigation
TRON, 17 f., 23
    3D surface tile, 20
    configuration, 18
    laboratory, 19
    schematic overview, 22
    simulation(s), 20

— V —

vertex, see vertices
Vision for Space Exploration, 13
Voyager 1, 13

— X —

XML meta information file, 80

www.ingramcontent.com/pod-product-compliance
Lightning Source LLC
Chambersburg PA
CBHW082203220526
45470CB00010B/3025